Acknowledgements

We would like to thank all those who helped with this research.

The Nutrition Advisory Committee of the Coronary Prevention Group advised on its aims and objectives. The members of this committee were Philip James (Chair), Wendy Doyle, Anne Foster, Walter Hare, Mike Lean, Suzanne May, Maggie Sanderson, Aubrey Sheiham, Tilo Ulbricht, Carol Williams and Jack Winkler.

Subject recruitment was carried out by USP Resources and Independent Research Bureau. John Bankart and Christopher Ecclestone helped with testing subjects. Christopher Raven of Information Design Unit, and Sarah Chapman and Alison Ford of the Em Space assisted in the preparation of materials.

David Buss, Anne Heughan, Janet Lewis, Elizabeth Morris and Maggie Sandersen, commented on drafts of the report. Ann Black, Simon Jacobs, Lucy Harris, Helen Lightowler, Jeanette Longfield, Michael O'Connor, Muriel Varley and Janet White also assisted in various ways.

We are particularly grateful to the Ministry of Agriculture Fisheries, and Food for funding the research.

Summary

This report describes the results of a two-phase research project into people's understanding of nutrition labelling. It was funded by the Ministry of Agriculture Fisheries and Food. The research was motivated by a recent European Community Directive – the 1990 Nutrition Labelling Rules Directive – which prescribes the format in which nutrition labelling must be presented. The Directive states that nutrition information should be shown numerically, but it allows for the possibility of including graphic representations according to formats yet to be determined.

The aim of the research was to assess whether people could understand numeric nutrition information and whether adding words or graphic information in various formats would help them to choose a healthier diet.

Phase 1 of the project comprised a series of group discussions. Participants were asked about their use of, and preference for, different nutrition labelling formats. Phase 2 consisted of a series of experimental studies in which participants were tested to see which formats were most useful in helping them make decisions about foods.

Results of phase 1: group discussions.
(For further details see pages 3-9.)

Most participants said they found numeric nutrition labelling difficult to understand because they had insufficient background knowledge to interpret the information given in this form. They said they would prefer an alternative scheme where nutrient levels were expressed using the words 'high', 'medium' and 'low'.

When asked to make judgements about single foods using numeric information, participants regularly made inaccurate assessments. This was because they did not know whether the figures represented high or low nutrient levels in relation to medical recommendations. Participants only used numeric information successfully when comparing foods where differences were simple and obvious.

Results of phase 2: experimental studies.
(For further details pages 67-78.)

These studies examined how adding words or graphic information affected participants' speed and accuracy when making judgements about single foods, or when comparing two or more foods.

Verbal banding, where words such as 'high', 'medium', and 'low' were used to indicate nutrient levels was the most consistently helpful format tested. (Examples of formats which were tested are shown on pages 67-75).

Graphic formats were found to be helpful to some people when they were direct representations of the nutrient levels. In particular supplementing verbal banding with bars (pages 70-71) where the nutrient level was directly related to the length of the bar helped participants who said they did not regularly use nutrition information on food packets.

However graphic formats which attempted to evaluate the nutrient level tended to confuse people. For instance a format which gave two stars for low fat and no stars for low fibre (page 73) caused people to make errors in their judgements about foods. The participants were confused by the inverse relationship between numbers of stars and the fat content as opposed to the direct relationship between the number of stars and the fibre content.

Dietary reference values (DRVs) indicating the recommended daily intake of each nutrient, and representations of the amount of a DRV supplied by a single serving, both helped some participants in their decision making. This was particularly the case for participants who were more interested in nutrition issues, and when the proportion of the DRV supplied by a serving was presented graphically (page 75).

Other findings emerged from this phase of the study. In particular, it was observed that extra information on the packet, such as health and nutrition claims, can affect people's understanding of nutrition labelling, whatever the format. In addition, people found it difficult to learn how to use more than one format for nutrition labelling.

Conclusion

This study provides evidence for the view that supplementing numeric nutrition information with words or well-designed graphic information can help consumers chose a healthier diet. Banding systems using words or graphic representations are more versatile and generally more helpful than formats based on DRVS.

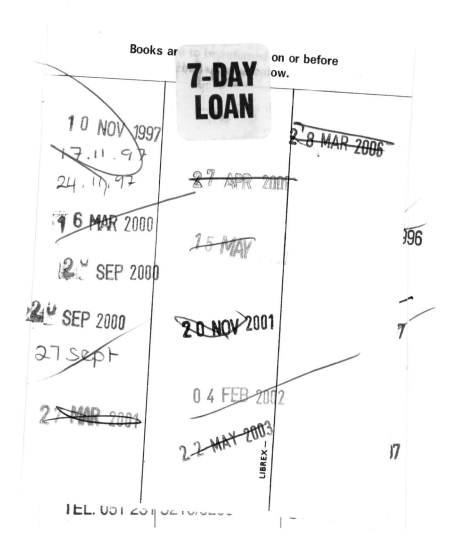

LONDON: HMSO

Alison Black is a psychologist and an independent researcher in cognitive
ergonomics and information design. **Michael Rayner** is the Senior
Research Officer for the Coronary Prevention Group.

Phase 1: # Contents

Phase 2: # Contents

Phase 1: Report of group discussions

1 Background to the study

The aim of this, discussion group, phase of the project was to find out what understanding of nutrition and other issues consumers brought to bear on decision-making about food. We took, as a starting point, one of the numerical formats for nutrition labelling that can be presented following the EC Nutrition Labelling Rules Directive* (see Figure 1), and investigated how consumers might use that format. The Directive provides for the presentation of nutrition information in graphic format, in addition to a numeric listing. So supplementary verbal and graphic representations of the numeric nutrition information were devised, and their effectiveness was explored in the context of tasks that consumers are likely to carry out when making decisions about foods to use or buy.

	per 100g	per serving (140g)
Energy	1487kJ/355kcal	2081kJ/497kcal
Protein	9.4g	13.2g
Carbohydrate	22.8g	31.9g
of which sugars	0.5g	0.7g
Fat	26.0g	36.4g
of which saturated fats	11.0g	15.4g
Dietary fibre	0.0g	0.0g
Sodium	0.7g	1.0g

Figure 1: One of the numeric nutrition information formats that can be presented following the EC Directive on Nutrition Labelling Rules. Energy is expressed in kilojoules and kilocalories, and seven specific nutrients are listed. The presentation of nutrition information per 100g is required. The additional presentation of nutrition information for an average serving is optional.

We were aware that exploring the use of nutrition information through group discussions was more likely to give a picture of people's declared comprehension or concern with nutrition issues, rather than their actual comprehension. So we included a series of questionnaire tasks within the discussions. Participants carried out the questionnaire tasks individually and then discussed them as a group. The questionnaire tasks provided us with individual profiles of participants' comprehension and likely use of nutrition information, which complemented the information gathered in the discussions.

We should emphasise here that discussion groups cannot provide a comprehensive view of the population's information needs, and we were not attempting to find out the level of comprehension or information needs of particular social classes. When we selected participants for the discussion groups, we aimed to include people for whom

---

* Commission of the European Communities (1990). *Council Directive of 24 September 1990 on nutrition labelling for foodstuffs.* 90/496/EEC.

nutrition information on food packaging might be helpful – people who were the prime decision-makers about food shopping in their household and who were aware of links between nutrition and coronary heart disease. We also aimed to include people with different levels of awareness of nutrition issues in order to make sure that our conclusions were not biased towards people with either more or less prior information about nutrition and health.

This first phase provided background for the development of sets of numeric, verbal and graphic representations of nutrition information which are tailored to consumers' understanding and needs. In the second phase these representations were tested with individual consumers, in decision-making tasks that resembled decisions made in shopping or planning meals.

2 | Summary of main findings of group discussions

Declared use of nutrition labelling
(see sub-section 4.4)

Most participants in the study claimed they currently looked at nutrition labelling on foods, although some said they looked only occasionally. Participants with both relatively high and standard levels of interest in nutrition and health claimed they looked for information about calories, sugar, fat, salt, fibre (and also additives). The predominant factor looked for was fat content. The more interested participants claimed they looked for several different factors; standard participants usually looked only for a single factor.

Use of weight information in nutrition labelling
(see sub-sections 6.1, 6.2 & 6.3)

Participants found it difficult to estimate what 100g portions of foods looked like. Additionally their estimates of the amounts of foods that would constitute a serving for themselves differed considerably for some foods. They said they did not use the weight information given in nutrition labelling (some were put off by the use of grams rather than ounces; many felt that manufacturers' recommended serving sizes were not appropriate). Health considerations usually influenced decisions whether or not to buy or eat a particular food, but not how much of that food to eat.

Declared comprehension of nutrition terms
(see sub-section 7.1)

At least 50% of the interested and standard participants claimed that they fully understood and were influenced in decision-making by the protein, carbohydrate, sugar and fat content of food. In addition, at least 50% of the interested participants also claimed they fully understood and were influenced by energy, kilocalories, saturated fat and dietary fibre. Saturated fat was less well understood and influential than fat. Sodium and kilojoules were poorly understood, and not influential.

Reception comprehension of nutrition terms
(see sub-sections 7.2 & 7.4)

Interested and standard participants showed similar, moderate levels of performance in a multiple choice task, despite the interested participants' declarations of their comprehension of more nutrition terms than standard participants. Performance was best for the terms: fat, saturated fat, saturates and salt. Subsequent discussion of the task (7.4) showed that even though participants answered the multiple choice questions successfully there was confusion about the relationship between saturated fat and fat, and the role of sodium in a healthy diet.

Production comprehension task
(see sub-section 7.3)

Participants had difficulty explaining carbohydrate/sugar and fat/saturated fat inclusion relationships, even though many had noticed the typographic signalling of these relationships in nutrition labelling. The interested participants were only slightly better at explaining the relationships than the standard participants.

Participants recommendations for eating for a healthy heart
(see sub-section 7.4)

The recommendations of both interested and standard participants tended to be simple maxims to avoid specific foods or general recommendations for 'balance'. They did not display particular appreciation of the role of different nutrients in the diet.

Using numeric information to assess single foods
(see sub-section 8.1)

Both interested and standard participants made inaccurate assessments of the levels of nutrients in single foods because they did not know what levels of nutrients might be appropriate to look for.

Using numeric information to compare foods
(see sub-section 8.2)

Participants could make simple comparisons between foods using numeric information, but not complex comparisons where, for example, a higher level of fat in one food had to be offset against a lower level of sugar. They tended to use a single nutrient level (usually fat, probably because of the bias of the discussion groups) as a yardstick to assess the food as a whole.

Attitudes to using numeric nutrition information
(see sub-section 8.3)

Some participants found numeric information difficult to use. Many felt they had insufficient background knowledge to understand the numeric information. Some participants suggested that information about the recommended daily intake of nutrients should be presented as a supplement to help interpret numeric information. Others thought direct health warnings about the levels of particular nutrients would be useful.

Preferences for selected alternatives to numeric information
(see sub-sections 9.1 & 9.2)

From a range of formats for the presentation of nutrition information (see Figure 2 on pp. 5–6) the preferred format for most decision-making tasks was a banding scheme, used selectively for nutrients felt to have a bearing on health. Some participants preferred a banding scheme for all nutrients, and others preferred a banding scheme supplemented by bars to give an immediate graphic emphasis to the levels of nutrients in foods. Participants felt their understanding of nutrition was sufficient for them not to be confused by different mappings in banding schemes, for example in bandings for fat and fibre, where 'high' is bad for fat but good for fibre. Questions were raised about how banding schemes were worked out, and what authority they might have.

Many interested participants gave high ratings to formats showing the percentage of recommended daily allowances provided by a serving of the food. Ratings for percentages of recommended daily allowances were highest for tasks where participants felt there would be plenty of time to make decisions.

a *Numeric listing*

	per 100 grams	per serving (330g)	
Energy	659	2175	kJ
	157	518	kcal
Protein	7.8	25.7	g
Carbohydrate	10.2	33.7	g
of which sugars	1.1	3.6	g
Fat	10.0	33.0	g
of which saturated fats	4.7	15.5	g
Dietary fibre	3.0	9.9	g
Sodium	0.32	1.1	g

b *Bar chart*

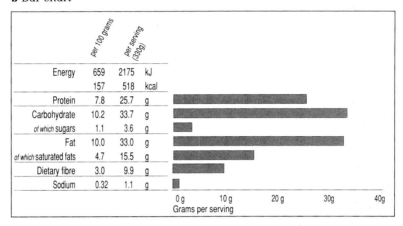

	per 100 grams	per serving (330g)	
Energy	659	2175	kJ
	157	518	kcal
Protein	7.8	25.7	g
Carbohydrate	10.2	33.7	g
of which sugars	1.1	3.6	g
Fat	10.0	33.0	g
of which saturated fats	4.7	15.5	g
Dietary fibre	3.0	9.9	g
Sodium	0.32	1.1	g

0 g   10 g   20 g   30g   40g
Grams per serving

c *Banding*

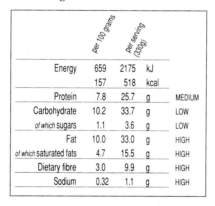

	per 100 grams	per serving (330g)		
Energy	659	2175	kJ	
	157	518	kcal	
Protein	7.8	25.7	g	MEDIUM
Carbohydrate	10.2	33.7	g	LOW
of which sugars	1.1	3.6	g	LOW
Fat	10.0	33.0	g	HIGH
of which saturated fats	4.7	15.5	g	HIGH
Dietary fibre	3.0	9.9	g	HIGH
Sodium	0.32	1.1	g	HIGH

d *Banding with bars*

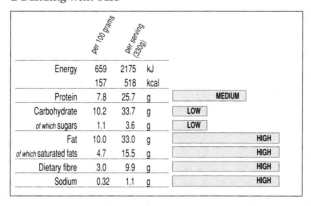

	per 100 grams	per serving (330g)		
Energy	659	2175	kJ	
	157	518	kcal	
Protein	7.8	25.7	g	MEDIUM
Carbohydrate	10.2	33.7	g	LOW
of which sugars	1.1	3.6	g	LOW
Fat	10.0	33.0	g	HIGH
of which saturated fats	4.7	15.5	g	HIGH
Dietary fibre	3.0	9.9	g	HIGH
Sodium	0.32	1.1	g	HIGH

Figure 2: *Alternative formats for nutrition information*
(Figure 2 continues on page 6)

e *Listing of RDA*

	per 100 grams	per serving (330g)		Recommended daily allowance
Energy	659	2175	kJ	
	157	518	kcal	
Protein	7.8	25.7	g	75g
Carbohydrate	10.2	33.7	g	345g
of which sugars	1.1	3.6	g	72g
Fat	10.0	33.0	g	80g
of which saturated fats	4.7	15.5	g	27g
Dietary fibre	3.0	9.9	g	30g
Sodium	0.32	1.1	g	2g

f *Percentage of RDA*

	per 100 grams	per serving (330g)		Percentage of recommended daily allowance (per serving)
Energy	659	2175	kJ	
	157	518	kcal	
Protein	7.8	25.7	g	34%
Carbohydrate	10.2	33.7	g	10%
of which sugars	1.1	3.6	g	5%
Fat	10.0	33.0	g	41%
of which saturated fats	4.7	15.5	g	57%
Dietary fibre	3.0	9.9	g	33%
Sodium	0.32	1.1	g	55%

g *Percentage RDA with bars*

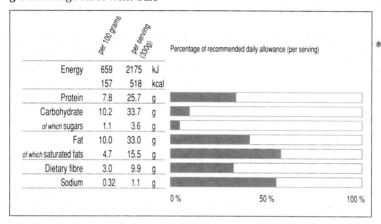

	per 100 grams	per serving (330g)		Percentage of recommended daily allowance (per serving)
Energy	659	2175	kJ	
	157	518	kcal	
Protein	7.8	25.7	g	
Carbohydrate	10.2	33.7	g	
of which sugars	1.1	3.6	g	
Fat	10.0	33.0	g	
of which saturated fats	4.7	15.5	g	
Dietary fibre	3.0	9.9	g	
Sodium	0.32	1.1	g	

0 %   50 %   100 %

h *'Plus points'*

	per 100 grams	per serving (330g)	
Energy	659	2175	kJ
	157	518	kcal
Protein	7.8	25.7	g
Carbohydrate	10.2	33.7	g
of which sugars	1.1	3.6	g
Fat	10.0	33.0	g
of which saturated fats	4.7	15.5	g
Dietary fibre	3.0	9.9	g
Sodium	0.32	1.1	g

✓ **Low in sugars**
High in dietary fibre

i *Selective banding*

	per 100 grams	per serving (330g)	
Energy	659	2175	kJ
	157	518	kcal
Protein	7.8	25.7	g
Carbohydrate	10.2	33.7	g
of which sugars	1.1	3.6	g
Fat	10.0	33.0	g
of which saturated fats	4.7	15.5	g
Dietary fibre	3.0	9.9	g
Sodium	0.32	1.1	g

Sugars	*Low*
Fat	*High*
Saturated fats	*High*
Sodium	*High*
Dietary fibre	*High*

Figure 2 (continued): *Alternative formats for nutrition information*

3 Conclusions from group discussions

3.1 Using nutrition information in decision-making

The combined group discussions and questionnaire responses helped us build up a preliminary picture of peoples' understanding of nutrition labelling, of how that understanding is brought to bear on decision-making in shopping and planning meals, and of other information sources that are drawn upon as specific decisions are made.

The results show that people are more likely to draw on their own background knowledge, of nutrition, rather than information presented explicitly on food packaging. This is not at all surprising considering the busy schedules that most people have – many participants in this study emphasised that most of their shopping is done under time pressure and that there is rarely time to make detailed consultations of nutrition information.

The discussion groups showed that much of the background knowledge that people do draw on is vague, especially the knowledge that they claim they use to ensure they have a balanced diet (see section 7, on people's comprehension of and response to nutrition factors). Some may have more useful background information in the form of strategies to avoid certain foods that are linked with either heart disease or cancer (see sub-section 7.4). However, this strategic information may not be very helpful when they are making decisions about processed foods, such as prepared meals, that are not covered by maxims such as 'eat less animal fat'.

People do look at explicit information on packaging when they are concerned about following a special diet. In most cases they are looking at single factors, such as calorie content, presence or absence of certain nutrients (such as fat, sugar or salt), or additives. They may also look at the packaging to find out the weight of the contents for price comparisons. If they are looking for more than one factor they may look for information in different places on the packaging.

Generally people appear to process the nutrition information currently available on food packaging at a superficial level. If they are looking at nutrition information they do not involve themselves with the numeric information to any degree, unless they have specific health reasons for doing so (see sub-section 6.3). They do not make numerical comparisons between the amounts of nutrients in different foods (see sub-section 8.3) and, if they make comparisons between the levels of nutrients in single foods, these may well be incorrect because the comparisons may be based on assumptions that quantities of all nutrients can be evaluated on the same scale (see sub-sections 8.3 and 9.1).

3.2 Users with different levels of interest in nutrition and health

Many consumers choose not to prioritise health considerations when they are making decisions about food (see sub-section 6.3). Others feel there is not enough time to think about health issues. Others may be intimidated by the amount and technical appearance of nutrition information presented to them, especially when it is presented in metric units (see sub-sections 8.3 and 9.1), or may not be comfortable with the kinds of calculations necessary to be able to make use of that information.

Differences in the comprehension for people with different levels of interest in nutrition were not obvious in the reception comprehension tasks described in sub-section 7.2 (these kinds of tasks are commonly used to assess people's comprehension of nutrition issues in question-naire studies). But differences became apparent in more complex production tasks (described in sub-section 7.3).

Even though it was possible to isolate a sub-group of consumers who had a more sophisticated understanding of nutrition issues, and might well be interested in more detailed nutrition information than other consumers, the differences between the two sub-groups should not be exaggerated. Many of the more interested participants still had miscon-ceptions and confusions about aspects of nutrition and health.

3.3 Matching nutrition information to tasks and users

Nutrition information could be tailored both to the tasks consumers carry out when making decisions about foods, and to the needs of consumers with different levels of interest in nutrition. It could be an immediately accessible and comprehensible representation of nutrient content, presenting information relating to a range of health issues (consumers following very specific diets might still need to go beyond this immediate presentation to more detailed numeric information). The information would then help consumers place individual foods within the context of their daily requirement for specific nutrients as they plan meals or make decisions about portion sizes.

Immediate information access via a banding system

There were calls for information saying 'what's good and what's bad' in foods, and for the equivalent of government health warnings on cigarettes, but food manufacturers are unlikely to adopt such a practice. A banding system, either for all nutrients or for nutrients considered relevant to health, would be a more suitable alternative. Such a system would give a 'first line' of information, preceding numeric information. Banding systems were tested out in the discussions, and all the systems tested (verbal formats or combined verbal and graphic formats) were popular with all groups, although some participants thought that graphic representation of the banding system was unnecessary.

Recommended daily allowance information for detailed consideration

There were additional calls for contextual information that would help consumers relate the numeric information for individual foods to daily requirements. Systems showing servings in relation to recommended daily allowances (RDAs) (in numeric and graphic formats) were presented to the discussion groups. Although some (more interested) participants saw their value, many found them difficult to understand, and even intimidating. It's worth bearing in mind that in the groups the RDA information was shown as the main source of information comple-menting standard numeric displays. It may be that, if it is used in combination with a banding system, which prepares users for the information they may need to interpret, it could be more successful. RDAs would be unlikely to be used by all consumers, but could be available for those who were more interested to draw on when required.

3.4 Following up phase 1 of the project

Our next step was to conduct detailed testing of individual subjects carrying out decision-making tasks using different formats of nutrition information (phase 2 of the project). Following the findings of phase 1 we decided to focus on banding systems and representations of RDAs.

The discussion groups raised some specific questions about the formats used for banding systems: whether they should be selective or comprehensive, whether they should be verbal or combine verbal and graphic elements. Further questions could be asked about how people's efficiency in using banding systems might be affected by the number of levels of bands in the system – the systems examined in this phase had three bands (high, medium, low), but both two- and four-level systems could be investigated.

There were also questions about how RDAs should be represented: the graphic system used in the discussion groups was not popular, and a suggestion was made for a unit based system (dividing RDAs into ten units). We decided to compare this kind of representation with listings of percentages of RDAs.

4 The participants in the group discussions

Five groups were recruited: two in Reading and three in Manchester. All participants were women of social class C1 or C2, who claimed to be the prime decision-maker about shopping, and the prime shopper in their household. Their ages ranged from 18-66. Men were not excluded deliberately from the study, but were eliminated when it proved difficult to recruit men who were both the prime decision-makers about shopping and met other criteria for inclusion in the study (see sub-section 4.1). The recruitment questionnaire for the groups is shown in Appendix 1 on p. 49.

4.1 Standard and interested groups

We had planned to talk to groups of people who had different levels of awareness about the links between nutrition and coronary heart disease, aiming to screen out the different groups by the recruitment question, 'Are you aware of any link between the food you eat and the chance of developing coronary heart disease?'. In fact none of the people interviewed said they were not aware of any link, and so we separated respondents into groups with different levels of interest in nutrition issues on the basis of their responses to the question 'Can you tell me the main things you need to do to eat for a healthy heart?'. People whose responses included no more than one of the five points A, B, C, D, E listed in Figure 3 were assigned to the standard groups. People who included three or more of the five points listed in Figure 3, including point E regarding salt, and who used the terms 'saturated' or 'polyunsaturated' in their responses were assigned to the interested groups.

Can you tell me the main things you need to do to eat for a healthy heart?

Check for:

A – change to reduced fat milk

B – substitute polyunsaturated margarine for butter or other animal fats

C – reduce fatty red meats, substitute chicken/fish

D – reduce overall consumption of full fat dairy products

E – reduce salt (sodium) consumption

Figure 3: Screening question used to distinguish standard and interested groups during recruitment.

There was one standard group in Reading (RS group) comprising six people and two standard groups in Manchester (MS_1 and MS_2 groups) each comprising five people. There was one interested group in Reading (RI group) comprising seven people and one interested group in Manchester (MI group) comprising six people.

4.2 Individual profiles of group participants

We gathered additional information about the participants' interest in nutrition at the focus groups, where participants answered a profile section in the questionnaire (the profile section is shown in Figure 4; full questionnaire is shown in Appendix 2 on pp. 50-57).

- Age
- School leaving age
- Qualifications, if any, gained since leaving school
- If following any special diet
- Details of special diet (*fixed choice response*)
  vegetarian   slimming   religious   low-fat   other
- Frequency of weighing foods (in cooking) (*fixed choice response*)
  at least once a week   at least once a month
  at least once a year   never
- Rating on a scale of 1 (low) – 7 (high) of interest in cooking
- Rate on a scale of 1 (low) – 7 (high) of interest in healthy eating
- Frequency of examining nutrition information on food packaging (*fixed choice response*)
  never   occasionally   frequently   always
- Occasions when nutrition information is looked at (*fixed choice response*)
  when shopping   when planning meals at home   other
- Reasons why nutrition information is looked at

Figure 4: Summary of personal profile questions completed by focus group participants before discussion sessions (the full questionnaire is shown in Appendix 2 on pp. 50-57).

The responses suggested that the standard and interested groups were not as distinct as might have been expected. It was likely that the recruitment procedure separated out groups who were more or less confident about their understanding of nutrition issues (and so were more or less happy to make recommendations about eating for a healthy heart to the interviewer during recruitment), but who had a range of levels of comprehension.

4.3 QI and QS sub-groups

The questionnaire allowed us to isolate individual participants within the groups who had a higher level of interest in nutrition than the other participants: people who claimed relatively high levels of interest in cooking and healthy eating (scoring joint ratings of 10 or more) and said that they always or frequently looked at nutrition labelling. In some parts of the following discussion the responses of these people are separated out from those of other participants. The two sub-groups with different levels of interest have been labelled QI (questionnaire interested) and QS (questionnaire standard). The QI sub-group comprised 13 people and the QS sub-group 16 people. The two sub-groups are shown in Table 1 on pp. 12-13.

Although the numbers in this study were small it was possible to find trends that suggested that the QI and QS sub-groups differed in ways other than their interest in nutrition issues. The QI sub-group tended to be younger than the QS sub-group – 10 (77%) of the QI sub-group were under 45 years old compared to five (31%) of the QS sub-group. Although the school leaving ages of the sub-groups were similar, the QI

Table 1: QI sub-group – Reading, standard (RS) and interested (RI), Manchester standard (MS$_1$ and MS$_2$) and interested (MI) groups

	Age	School leaving age and qualifications since leaving school	Special diet	Frequency of weighing food	Interest in cooking (1-7)	Interest in healthy eating (1-7)	How often nutrition info. looked at	When information looked at	Reason for looking at nutrition information
RS2	25-35	16, 2 A-levels/ACII	none	at least once a week	7	7	frequently	shopping and meal planning	fat and sugars, calorie content
RS3	36-45	18, teaching certificate	slimming, low-fat	at least once a week	4	7	frequently	shopping and meal planning	calories, low fat
RS6	26-35	16, none	slimming	at least once a week	4	6	frequently	shopping	calories, fat content
RI1	26-35	16, certificate of office studies	low fat	never	5	5	frequently	shopping and meal planning	energy value, then fat content then sugar
RI4	46-55	17, none	none	at least once a week	3	7	always	shopping	salt, fat and sugar content
RI5	26-35	16, City and Guilds catering	slimming, low fat	at least once a week	6	7	frequently	shopping	fat, sugar, colour content
MS$_1$1	56-65	15, none	none	at least once a week	7	5	frequently	shopping	fat, sugar, fibre content
MS$_1$2	46-55	15, none	slimming, low-fat	at least once a week	7	7	always	shopping and meal planning	fat and sugar content, calorie value, fibre content
MS$_2$4	18-25	18, O-levels, nursing care certificate	none	never	5	6	frequently	shopping	calories, fat, sugar content
MI1	36-45	17, none	diabetic	at least once a month	5	6	always	shopping	carbohydrate level, fat
MI3	18-25	16, none	low cholesterol	at least once a week	6	7	always	shopping	calories, fat, fibre
MI4	26-35	16, S.E.N.	slimming	at least once a week	4	6	frequently	shopping	calories, fat, fibre
MI5	36-45	18, City and Guilds	none	at least once a week	7	7	frequently	shopping	fat, carbohydrate, fibre

Table 1 continues on page 13

Table 1 (ctd): QS sub-group – Reading, standard (RS) and interested (RI), Manchester standard (MS$_1$ and MS$_2$) and interested (MI) groups

	Age	School leaving age and qualifications since leaving school	Special diet	Frequency of weighing food	Interest in cooking (1-7)	Interest in healthy eating (1-7)	How often nutrition info. looked at	When information looked at	Reason for looking at nutrition information
RS1	56-65	16, no further qualifications	religious	at least once a week	4	4	occasionally	shopping	E numbers
RS4	46-55	15, cook supervisor	none	never	7	4	never	–	–
RS5	46-55	17, no further qualifications	none	at least once a week	4	4	never	–	–
RI2	46-55	15, no further qualifications	slimming	never	1	3	occasionally	shopping	low fat
RI3	56-65	14, no further qualifications	low fat	at least once a week	4	5	occasionally	shopping	fat, fibre
RI6	26-35	17, no further qualifications	none	at least once a week	3	7	frequently	shopping	calories, salt, additives etc.
RI7	36-45	18, no further qualifications	none	at least once a week	5	5	occasionally	if interested in contents	what the product contains
MS$_1$3	over 65	14, no further qualifications	none	at least once a week	4	5	occasionally	if cooking for vegetarians	animal products
MS$_1$4	18-25	18, no further qualifications	none	never	4	5	occasionally	–	sugars, salt, calories
MS$_1$5	46-55	18, no further qualifications	none	never	4	6	occasionally	shopping	fat content
MS$_2$1	46-55	15, none	none	never	4	4	occasionally	shopping	if high in fat
MS$_2$2	56-65	14, no further qualifications	none	never	1	1	never	–	low fat
MS$_2$3	46-55	15, no further qualifications	low fat	at least once a week	7	7	occasionally	shopping	fat and sugar content
MS$_2$5	46-55	15, O.T.C. first aid and home nursing	none	never	4	4	occasionally	shopping	low in cholesterol, low in polyunsaturates
M12	36-45	16, RSA Info. tech	none	at least once a week	6	2	always	shopping	nutrition, value for money
M16	18-25	18, secretarial	none	never	4	4	frequently	shopping	calories, vitamins, additives

sub-group were more likely to have gained some qualification since leaving school. Many more of the QI than the QS sub-group were following some sort of diet (usually a slimming or low-fat diet) – eight (67%) of the QI sub-group compared to four (25%) of the QS sub-group. Possibly as a consequence of following a slimming diet the QI sub-group tended to weigh foods more than the QS sub-group. The QI sub-group were more likely to look at nutrition information in meal planning as well as shopping, whereas those in the QS sub-group who consulted nutrition information only did so when shopping. This may also be because there were more dieters in the QI sub-group.

4.4 The QI and QS sub-groups' current use of nutrition labelling

The range of reasons for consulting nutrition information was similar across the two sub-groups: people were looking for information about calories, sugar, fat, salt, fibre, or additives. And in both sub-groups the predominant reason for looking at nutrition information was to find out about fat content, with few people looking for information about salt and fibre (these data will be discussed in more detail in sub-section 7.1). But as Table 2 (on p. 15) shows, people in the QI sub-group were more likely to be looking for several different kinds of nutrition information. As well as fat content they tended to look for information about calories and sugar. In contrast, people in the QS sub-group who looked at nutrition information were more likely to be looking for less than two factors. These data suggest that people in the QI sub-group may attempt a more complex reasoning task. Possibly they were more aware of the complex interactions between nutrition and health, or, generally, more adept at carrying out reasoning tasks.

Table 2: Reasons for looking at nutrition information

Questionnaire standard group (QS)

	Calories	Sugar	Fat	Salt	Fibre	Additives
RS1	0	0	0	0	0	1
RS4	–	–	–	–	–	–
RS5	–	–	–	–	–	–
RI2	0	0	1	0	0	0
RI3	0	0	1	0	1	1
RI6	1	0	0	0	0	0
RI7	–	–	–	–	–	–
MS$_1$3	–	–	–	–	–	–
MS$_1$4	0	1	0	1	0	0
MS$_1$5	0	0	1	0	0	0
MS$_2$1	0	0	1	0	0	0
MS$_2$2	0	0	1	0	0	0
MS$_2$3	0	1	1	0	0	0
MS$_2$5	0	0	1	0	0	0
MI2	–	–	–	–	–	–
MI6	1	0	0	0	0	0*

Questionnaire interested group (QI)

	Calories	Sugar	Fat	Salt	Fibre	Additives
RS2	1	1	1	0	0	0
RS3	1	0	1	0	0	0
RS6	1	0	1	0	0	0
RI1	1	1	1	0	0	0
RI4	0	1	1	1	0	0
RI5	0	1	1	0	0	1
MS$_1$1	0	1	1	0	1	0
MS$_1$2	1	1	1	0	1	0
MS$_2$4	1	1	1	0	0	0
MI1	0	1†	1	0	0	0
MI3	1	0	1	0	1	0
MI4	1	0	1	0	1	0
MI5	0	1†	1	0	1	0

*participant also wrote 'vitamins'
†participant also wrote 'carbohydrate'

15

5 Conducting the group discussions

The questionnaire and group discussion were planned to form integrated sessions. Participants filled out the personal profile section of the questionnaire at the beginning of the discussion (the complete questionnaire is shown in Appendix 2). Thereafter they filled in further sections of the questionnaire only when they were asked to do so by the discussion moderator. So participants were all kept at the same place in the questionnaire.

After the groups had completed each section of the questionnaire the moderator directed a discussion of the issues addressed in the questionnaire, and participants could refer back to their questionnaire responses as they contributed to the discussion.

The discussion of different formats for presenting nutrition information departed from the order of questionnaire first, followed by discussion. Here the groups discussed the nutrition information formats together before they gave them individual ratings in their questionnaires. This change of order was necessary to stimulate participants to think about the functional consequences of using different formats. It seemed likely that participants would not have thought about different ways of presenting nutrition information before the session. So asking them to evaluate different formats without any prior discussion could elicit superficial rather than considered responses.

6 Judging food weights and portion sizes

6.1 100 gram task

Aim

Since the EC Directive specifies that nutrient content should be expressed per 100g, we aimed to find out how well participants could judge what 100g of various foods looked or felt like. Participants were asked to examine (look at and pick up) three different size samples of a range of foods and decide which one weighed 100g. The subsequent discussion covered how, more generally, people thought about the quantities of food they bought.

The task

Nine foods were used in three groupings: three 'generic' foods (apples, butter, potatoes), three processed foods (baked beans, beef-burgers and marmalade) and three prepared dishes (fish in cheese sauce, chocolate gateau, pizza). Within each grouping, the 100g portion was the smallest, medium or largest portion for one of each of the foods. The RS group carried out this task without being told the ounce equivalent of 100g, but the other groups asked what the equivalent was, and so were told.

Questionnaire responses

The choices people made are shown in Table 3. (Note that the MI group did not carry out this task, and MS_15 misunderstood the task instructions so her responses are not included.)

The average number of correct choices out of a possible total of 9 was 4.3 (range 1 – 7). The QI (questionnaire interested) sub-group scored an average of 5 (range 3 – 7) compared to the QS (questionnaire standard) sub-group who scored an average of 3.7 (range 1 – 6). Pooling across the sub-groups, people who weighed foods more than once a week scored an average of 4.7 (range 2 – 7) whereas people who weighed foods less frequently scored an average of 3.5 (range 1 – 6). The scores for the QI sub-group and for the people who weighed foods more than once a week were significantly above chance ($p < .01$ and $p = .001$ respectively), but scores for the QS sub-group and people who weighed foods less than once a week were not.

We anticipated that people would find it easier to choose the 100g portion for the generic foods, which they might buy by weight or weigh in cooking, than the processed foods or prepared dishes, but this was not the case. The average correct scores (maximum three) for generic foods, processed foods and prepared courses were 1.6, 1.18 and 1.5 respectively.

Table 3: *Responses in 100 gram task for participants in the Reading Standard (RS), Reading Interested (RI), Manchester Standard (MS₁ and MS₂) groups*

1 – correct; 0 – incorrect. Participants whose questionnaire response suggested a higher level of interest in nutrition than other participants (QI group) are marked by grey shading

	apples	butter	potatoes	baked beans	beef-burgers	marmalade	fish in sauce	gateau	pizza	*Total*
RS1	0	0	1	0	0	0	1	1	1	4
RS2	1	1	1	0	0	1	1	1	1	7
RS3	0	0	0	1	0	0	0	0	1	2
RS4	0	0	0	0	0	1	0	1	1	3
RS5	0	0	0	0	0	1	0	0	1	2
RS6	1	0	1	1	0	0	1	1	1	6
RI1	1	0	1	0	0	1	1	1	1	6
RI2	0	1	0	0	1	0	0	0	0	2
RI3	1	0	0	1	0	1	1	0	1	5
RI4	1	1	1	0	0	1	0	0	1	5
RI5	0	1	1	0	0	1	0	0	1	4
RI6	1	0	0	1	0	0	0	0	1	3
RI7	0	1	0	1	0	0	0	1	1	5
MS₁1	1	1	1	1	1	0	0	1	0	6
MS₁2	1	0	1	1	1	0	1	1	0	6
MS₁3	1	1	0	1	1	1	0	1	0	6
MS₁4	1	0	1	0	0	1	1	0	0	4
MS₁5	–	–	–	–	–	–	–	–	–	–
MS₂1	1	1	0	1	0	0	0	1	1	5
MS₂2	0	1	0	0	0	0	0	0	0	1
MS₂3	1	0	1	0	0	1	1	1	0	5
MS₂4	0	1	1	0	1	0	0	0	0	3
MS₂5	1	0	1	0	1	1	0	0	0	4

6.2 Average portion task

Aim

In order to find out whether consumers are likely to be able to use nutrition information provided on a per serving basis, we set participants a task where they examined food samples and said whether they would normally eat more or less, or about the same as the given serving size.

The task

Samples of the three processed foods and three prepared courses used in the 100g task were made up either to the serving size given by the manufacturer in the nutrition labelling, or to an average portion size taken from Food Portion Sizes (1988).* The sample sizes were:

Baked beans – 135g; Beef-burgers – 35g (1 beef-burger); Marmalade – 15g; Fish in sauce – 170g; Gateau – 85g; Pizza – 215g (half a pizza).

People examined the samples and said whether they would normally eat more or less, or about the same size serving.

---

* Crawley, H. (1988). *Food portion sizes*. London: HMSO.

Questionnaire responses

As Table 4 shows, there was a stronger consensus about some foods than about others. People thought they would be likely to eat more than the single beef-burger, about the same as the gateau portion and less than the pizza serving, but there was less agreement about other foods.

6.3 Group discussion

100 gram task

Participants said they found this task hard. People who weighed foods regularly were surprised at its difficulty. Most participants had used the strategy of taking a known referent (such as four ounces of butter) to estimate the weight of other portions. When asked if the task would have been any easier had it been to judge four ounce portions rather than 100g portions, all groups said it would, even the three groups who had been told that 100g were 'just under four ounces.' Some younger participants (RS2 and MS24) said they were just as happy working in grams as in ounces, but other younger participants (RS6 and MS14) preferred ounces. Older participants were emphatic that they preferred ounces to grams.

When asked if using grams was a difficulty in shopping, most participants said they did not think it was. It was still possible to weigh loose foods in ounces and they only looked at the weight of packaged foods to make judgements about the relative value for money of different size packs. In this case working in grams was no different from working in ounces. (Although not for some older members of the groups who felt that, whatever the task, they could only 'think in ounces'.)

When asked if any foods were particularly difficult to judge, participants thought that processed foods, which they rarely weighed, were more difficult. However, the data do not suggest participants were more accurate with unprocessed foods, despite one participant's claims that she could judge fruit and vegetables 'spot on'.

Pizza, marmalade and gateau were singled out as particularly difficult to judge. Most people thought that all the pizza portions looked too large to be 100g and the marmalade too small. One participant in the RI group said that she thought gateau would be 'a surprise' in that a relatively large amount might be needed to make up a 100g portion. The discussions in all of the groups suggested that people did not really have any explicit understanding of the different densities of foods, except that they were aware that some foods, such as crisps or cornflakes, had a very low density.

None of the participants felt they ever carried out the kind of judgement tasks they had attempted here in the normal course of shopping. The practice of quoting nutrition information for 100g had made little impact on people who looked at nutrition information, since they were not focussing attention on relating nutrition information to weight. It was hard to tell whether they would do so even if they were better at 'thinking in grams' than they appeared to be here.

Average portion task

a. Factors influencing portion size

When participants were asked what normally influenced the size of portion they ate they said initially that the main influence was how hungry they were at the time. The ensuing discussion suggested that a range of mediating factors influenced their choice of portion size,

Table 4: *Responses in the average portion task for participants in the Reading Standard (RS), Reading Interested (RI), Manchester Standard (MS$_1$ and MS$_S$) and Manchester Interested (MI) groups*

L – would eat less; S – would eat the same; M – would eat more

	baked beans	beef-burgers	marmalade	fish in sauce	gateau	pizza
RS1	L	M	M	S	S	L
RS2	S	M	S	S	S	L
RS3	L	M	L	S	S	L
RS4	S	M	S	S	S	L
RS5	M	M	S	S	S	L
RS6	M	S	L	M	S	S
RI1	L	S	M	S	S	L
RI2	M	M	S	M	L	L
RI3	S	S	M	M	S	L
RI4	M	M	S	M	S	L
RI5	M	M	S	S	S	L
RI6	L	M	S	S	S	L
RI7	S	M	M	M	S	L
MS$_1$1	S	M	S	M	M	L
MS$_1$2	S	M	S	S	S	L
MS$_1$3	S	M	M	S	S	L
MS$_1$4	S	S	L	M	L	L
MS$_1$5	M	M	L	S	M	L
MS$_2$1	S	S	L	M	L	L
MS$_2$2	L	M	S	L	S	L
MS$_2$3	M	S	S	S	S	L
MS$_2$4	M	L	–	–	M	L
MS$_2$5	S	S	L	S	L	L
MI1	S	M	S	M	S	M
MI2	M	M	S	S	S	S
MI3	L	S	–	M	L	L
MI4	L	S	–	M	L	L
MI5	S	S	S	S	L	L
MI6	L	S	L	S	L	S

including calorie content (dieters said they would usually like to eat more than they allowed themselves), people's mood when they were eating and the amount of time to eat. Participants also made judgements about portion size for their families, based on what they thought each member of the family needed (RS6 and RI7 commented that manual workers needed to eat more than sedentary workers) and their knowledge of family members' different tastes and preferences.

Cost was also mentioned, although this was more likely to influence initial decisions about whether or not to buy a food than how much to eat. The influence of cost on portion size was only direct in the case of boil-in-the bag fish which people said was so expensive that they would only eat one portion, even though it was, as one participant said, 'piddling'.

b. The influence of health considerations on portion size

When asked specifically whether health considerations influenced the amount of particular foods that they ate, many said not (particularly in the RS and MS_1 and MS_2 groups). Those who claimed to be influenced by health considerations, were more likely to be influenced when they were buying food, rather than when they were deciding how much to eat – they bought skimmed milk and low-fat yoghurts rather than their full-fat equivalents. The only clear examples of portion control were not eating as much cream and sugar as they would like and compensating for eating too much of a food they considered unhealthy (such as gateau) by eating less of everything else. Some participants ($MS_2$3 and MI2) said that although they might know that it would be better for them to eat less of a particular food, they did not always stick to this.

When asked in detail whether the fat content of a food influenced how much they ate, most participants said it did not.

c. The influence of recommended serving size on portion size

During this discussion no one spontaneously mentioned using the manufacturers' recommended portion size as a guide to the portion size they ate. When asked explicitly whether they used manufacturers' recommended serving size, the MI group laughed, saying it was never enough. Packaged foods serving more than one person were divided up among the family, according to their individual eating habits. Participants said that, provided a food was not too expensive, they often bought more packets than the serving size would suggest they needed to feed the whole family.

7 Comprehension of nutrition terms

Aims

We used a series of tasks to find out whether people's perception of their understanding of nutrition terminology matched their actual understanding of the terminology, and whether there were particular aspects of nutrition that were better understood than others. The first task was a declaration task, in which people stated whether they understood or were influenced by particular nutritional aspects of foods. This was followed by a reception comprehension task, in which participants demonstrated their level of comprehension by judging the accuracy of a series of statements about nutrition and coronary heart disease. Subsequently there was a production comprehension task, in which people defined relationships between different nutrients. The tasks were then used to focus discussion on the aspects of nutrition and health that were more or less difficult to understand.

7.1 Declaration task

The task

Participants were presented with a list of the terms commonly used in nutrition labelling – energy, kilojoules, kilocalories, protein, carbohydrate, sugars, fat, saturated fat, dietary fibre, sodium. These are also the terms that will have to be used in labels following the implementation of the EC Directive. Participants were asked to rate whether they fully understood, roughly understood, or did not understand the terms. They were also asked to say whether or not their decisions to buy or use particular foods were influenced by the energy values or by individual nutrients, and to list other factors that might influence their decisions.

Questionnaire responses

Table 5 (on p. 23) summarises the responses for the QI (questionnaire interested) and QS (questionnaire standard) sub-groups. For each term the percentage of participants answering that they fully or roughly understood the term are shown on the left-hand side of the column, and the percentage who fully understood or roughly understood and felt their decision making was influenced by the nutrient are shown on the right-hand side of the column. Table 6 (on p. 24) rank orders the nutrients according to the percentage of participants in each sub-group declaring that they fully understood them and were influenced by them in their decision making.

The ordering of influential factors was roughly similar for both sub-groups. However for each nutrient more participants in the QI sub-group than in the QS sub-group gave ratings showing they understood and were influenced by the nutrient in decision making. Eight of the nutrition terms (protein, fat, carbohydrate, sugars, dietary fibre, energy, kilocalories, saturated fat) were rated as fully understood and influential by 50% or more of the QI sub-group. In contrast, only four terms (sugars, fat, carbohydrate, protein) were rated as fully understood and

Table 5: Nutrition terminology – summary

Declared comprehension and influence of nutrition terms on decision-making for QI (questionnaire interested) sub-group (n = 13) and QS (quesionnaire standard) sub-group (n = 16). The percentages declaring full or rough comprehension are shown on the left in each column and the percentages declaring they are influenced in decision-making are shown on the right.

	energy		kilojoules		kilocalories		protein		carbohydrate		sugars		fat		saturated fat		dietary fibre		sodium	
QI group																				
Fully	77%	54%	15%	15%	54%	54%	100%	92%	92%	85%	92%	85%	92%	92%	69%	54%	85%	77%	69%	46%
Roughly	23%	0%	31%	8%	23%	15%	0%	0%	8%	8%	8%	8%	8%	0%	31%	31%	8%	8%	31%	15%
QS group																				
Fully	69%	44%	6%	0%	25%	25%	50%	50%	56%	56%	87%	81%	81%	81%	38%	25%	38%	25%	31%	25%
Roughly	31%	13%	6%	0%	6%	0% +13%*	44%	38% +6%*	44%	13%	13%	13%	19%	19%	44%	13% +6%*	50%	38%	44%	13%

*The additional figures shown for the QS group are for participants who claimed not to understand a term, but to be influenced by it in decision-making.

Table 6: *Rank ordering of comprehension of nutrition factors*

Ranking according to percentage of participants of QI (questionnaire interested) and QS (questionnaire standard) sub-groups declaring that they fully understand individual terms and are influenced by them in decisions about foods to buy or use.

QI sub-group (n = 13)

rank	nutrition factor	% declaring comprehension and influence
1	Fat	92%
1	Protein	92%
3	Carbohydrate	85%
3	Sugars	85%
5	Dietary fibre	77%
6	Energy	54%
6	Kilocalories	54%
6	Saturated fat	54%
9	Sodium	46%
10	Kilojoules	15%

QS sub-group (n = 16)

rank	nutrition factor	% declaring comprehension and influence
1	Fat	81%
1	Sugars	81%
3	Carbohydrate	56%
4	Protein	50%
5	Energy	44%
6	Saturated fat	25%
6	Dietary fibre	25%
6	Kilocalories	25%
6	Sodium	25%
10	Kilojoules	0%

influential by the QS sub-group. This difference in claims about influential factors suggests that the QI sub-group may attempt more complex reasoning than the QS sub-group when decision-making about food, and complements the data given in Table 2 (see section 4.4). However people's claims that they are influenced by particular factors may not mean that they are genuinely influenced by these factors.

Both sub-groups claimed a markedly higher comprehension and influence of fat than saturated fat (92% in the QI sub-group and 81% in the QS sub-group claimed to understand fat; 54% in the QI sub-group and 25% in the QS sub-group claimed to understand saturated fat).

In both the QI or the QS sub-groups sodium had relatively low priority in decision-making. And kilojoules were poorly understood, and therefore had little influence in both sub-groups.

When participants were asked about additional factors that influenced their decisions to buy or use foods, many mentioned food additives (preservatives, colourings and flavourings).

7.2 Reception comprehension task

The task

We used a multiple choice task to examine whether people's assessment of their comprehension of nutrition terms was borne out by their actual comprehension. Participants were shown a series of statements relating to coronary heart disease, using the nutrition terms set out in the EC Directive on Nutrition Labelling Rules. For each statement participants had to respond with 'true', 'false', 'it makes no difference' or 'I don't know'. Participants were also asked to mark responses they found particularly difficult to decide on. (The statements are shown, with the expected responses, in Figure 5.) We added two additional terms ('saturates' and 'salt') to the EC Directive terms in order to see whether participants' responses to statements using these terms differed from responses to their recommended counterparts 'saturated fats' and 'sodium'.

It's probably good for the heart to eat more dietary fibre
True

It's probably good for the heart to eat more sodium
False

It's probably good for the heart to eat fewer saturates
True

It's probably good for the heart to eat more salt
False

It's probably good for the heart to eat food with more kilojoules
False or *It makes no difference*

It's probably good for the heart to eat less saturated fat
True

It's probably good for the heart to eat more fat
False

It's probably good for the heart to eat more carbohydrate
True or *False* or *It makes no difference*

It's probably good for the heart to eat food with more kilocalories
False or *It makes no difference*

It's probably good for the heart to eat more sugars
False

It's probably good for the heart to eat less protein
False

It's probably good for the heart to eat food with more energy
False or *It makes no difference*

Figure 5: *Statements used in reception task to test comprehension of nutrition terminology, shown here with expected responses.*

Questionnaire responses

We excluded statements which had more than one possible response (statements relating to energy, kilocalories, kilojoules and carbohydrate) from the analysis of participants' responses. However it is worth noting the particular pattern of scores for the statements about energy and kilocalories. Ten of the participants thought that more energy was good for the heart, three thought more energy was not good for the heart and ten thought that more energy made no difference. In contrast, only one thought that more kilocalories were good for the heart, twelve thought more kilocalories were not good for the heart and twelve thought kilocalories made no difference. This pattern suggests that participants did not understand the relationship between energy and kilocalories. Generally, they viewed energy as neutral or good in relation to coronary health, but calories as neutral or bad. (Only nine participants gave a response for kilojoules: five of the nine who responded thought that more kilojoules were not good for the heart.)

Participants in the QI sub-group gave a mean of 6.4 expected reponses (maximum 10) and participants in the QS sub-group gave a mean of 5.8 expected responses. There was no significant difference between the scores for the two sub-groups.

Table 7 summarises the number of expected responses given for each nutrient and the percentage of participants in each sub-group giving

Table 7: *Nutrition terminology – percentage of participants in QI and QS sub-groups giving expected responses*

	Total correct scores (n = 29)	Percentage correct in QI sub-group (n = 13)	Percentage correct in QS sub-group (n = 16)
protein	10	23%	44%
sugars	20	85%	56%
fat	22	100%	94%
saturated fat	27	100%	88%
saturates	26	100%	81%
fibre	18	62%	63%
sodium	20	77%	63%
salt	25	92%	81%

these responses. For both sub-groups agreement with the expected response was highest for statements about fat, saturated fat, saturates and sodium.

Agreement was lowest for 'protein', possibly because it required a response 'it makes no difference.' It may be that although people have general notions about nutrition and health (both the QI and QS sub-groups said that protein influenced their decisions about foods to buy and eat), they are not confident about the relationship between nutrition and specific health problems, such as coronary heart disease.

There were fewer expected responses for the 'sodium' statement than for the equivalent 'salt' statement, reflecting people's difficulty with the term 'sodium.' However there were similar levels of expected responses for the 'saturates' and 'saturated fat.'

7.3 Production comprehension task

The task

We presented participants with the nutrition information for a prepared meal (see Figure 6) and asked them to explain why the words 'sugars' and 'saturated fats' were preceded by the phrase 'of which.'

	per 100g
Energy	848 kJ/202kcal
Protein	5.1g
Carbohydrate	10.6g
of which sugars	1.0g
Fat	15.8g
of which saturated fat	1.6g
Dietary fibre	1.0g
Sodium	0.5g

Figure 6: *Nutrition information shown in production comprehension task where participants were asked to explain the use of 'of which' before 'sugars' and 'saturated fats.'*

Questionnaire responses

The responses to the 'of which' questions were scored by three independent judges, and were only scored as correct if at least two out of the three judges felt they demonstrated that the participant understood the inclusion relationships between carbohydrates and sugars, and between fat and saturated fats (so our use of the term 'correct', below, is to some extent subjective).

There were nine correct responses to the carbohydrate/sugar question (five in the QI sub-group and four in the QS sub-group) and eleven correct responses to the fat/saturated fat question (seven in the QI sub-group and four in the QS sub-group). Participants found it difficult to explain the inclusion relationships, and this difficulty suggests they may have overstated their comprehension of some of the nutrition terms in the declaration task. The QI sub-group were slightly better at explaining the inclusion relationships than the QS sub-group, but not as good as might have been expected from their claims for their interest in nutrition issues.

7.4 Group discussion

Discussion of reception comprehension task

Participants were asked if there were any statements they found particularly difficult to respond to in the reception comprehension task.

a. Kilojoules and kilocalories

The initial response in all groups was to ask for clarification of 'kilojoules' and 'kilocalories.' Some thought that kilocalories were the same as calories. When this was confirmed, others were surprised. A participant in the RS group continued to express doubt that a kilocalorie could be 'one, just one?' calorie, and a participant in the RI group thought the term expressed the number of calories per kilogram. When asked if they had noticed these terms on food labelling, some had, but did not know and had not found out what they meant. MI2 said that if she was looking for calorific value on food labelling she looked at both kilojoules and kilocalories then went by whichever was the smallest.

b. Carbohydrate

All groups asked about carbohydrate. There were conflicting views of carbohydrate – in different groups it was described as a provider of energy, it was identified with 'stodge', it was described as 'essential for a balanced diet.' The groups' comments reflected the questionnaire responses: that carbohydrate was something that should be taken into account in the diet, but for reasons that were not clearly understood. There was no clear view about whether carbohydrate had a bearing on heart disease or not.

c. Sugars

As well as general confusion about carbohydrate, there was specific confusion about sugars, which were seen as high in calories and therefore bad for the heart (the questionnaire responses suggested that most participants had taken this view). The RS group talked about there being different kinds of sugars, concluding that not all were 'bad for you.'

d. Fat and saturated fats

Individuals in all except the MI group asked for clarification about whether it was saturated fats or unsaturated fats that were bad for the heart. But other participants in the groups seemed confident about the distinction between the two kinds of fats. The responses to the comprehension tests suggest that people's understanding of the distinction may be a limited one, providing a strategy for making some decisions about foods (demonstrated in the correct responses in reception comprehension), but not the basis for more complex reasoning (since most participants failed to explain the fat/saturated fats inclusion).

e. Sodium

All groups asked about sodium and salt. There was confusion about the term 'sodium' (as the questionnaire responses suggest). Most participants did not seem to be aware of the connections between salt consumption and high blood pressure or heart disease.

f. Dietary fibre

The RI group and MS_1 group asked about dietary fibre. Participants believed it was good for the bowel but were not aware of the possibility of a link with coronary heart disease. This lack of awareness is also demonstrated in the moderate level of expected responses in the reception comprehension task (see Table 7, on p. 26).

Discussion of production comprehension task

Many in the RS, MS_1 and MS_2 groups said that they had not seen typographic structuring or phrases such as 'of which' used in nutrition labelling. Others, in all groups, had noticed these conventions but had not understood them. Participants in the MI group said their attention was directed more to the calorie content of foods, so they hadn't bothered too much with other details.

All groups found answering the questions difficult, and there was further discussion of carbohydrate, how sugars related to carbohydrate, and the relationship of saturated fats to fat.

Discussion of nutrition recommendations

The discussion of the two comprehension tasks was broadened into a general discussion in which the groups were asked to make recommendations for eating for a healthy heart. By this stage the groups' attention had been focussed on the issues involved through discussion of the questionnaire.

All groups recommended eating less fat and less salt. Fibre was mentioned by individuals in the RS and MS_1 groups, but not strongly confirmed by other members of the group. Sugar was mentioned in the MS_1 group.

When prompted further, the groups gave a range of specific recommendations:

RS: don't eat too much protein, as in eggs...there's a lot of fat in eggs

RI: eat more lean meat and oily fish, reduce animal fats

MS_1: eat less red meat, lots of kippers, not too many eggs, fry in vegetable oil and not lard because 'it's not as much fat, is it', eat more brown bread and greens.

MS_2: eat chicken or fish rather than red meat, cut out cheeses that are high in fat, use skimmed milk.

MI: cut down on dairy products, don't eat red meat, shellfish or eggs.

When asked if there were aspects of nutrition that were important for health other than for the heart the groups talked of 'balanced diets', 'fresh foods' (fruit and vegetables), 'natural foods.'

The responses to these questions perhaps reflect the view of nutrition and health suggested by the discussion of the reception and production comprehension tasks: people tend to work to basic maxims that they can apply (if they choose) during shopping and preparing meals, such as 'cut down on animal fats', 'cut down on sugar'. They are not relying on a deep understanding of nutrition/health relationships, even when they consider themselves to be particularly interested in food and healthy eating. The basic maxims are translated to strategies that allow them to make decisions relatively quickly and simply, such as 'cut down on dairy products', 'cut down on eggs', 'less red meat' and so on.

Judging and comparing
nutrient levels

Aim

We aimed to assess whether participants could make judgements about
the levels of nutrients (high, medium, low) on the basis of numerical
information in nutrition labelling, and to examine whether different
relationships among levels of nutrients influenced people's judgements
about single foods and comparisons between pairs of foods.

8.1 Assessing single foods

The task

Participants examined nutrition labelling for three prepared meals.
They examined each set of information singly, without reference to the
information for other meals, and in the order shown in Figure 7. The
participants had to judge whether each meal was high, medium or low
in sugars, fat, saturated fats, sodium and dietary fibre. The participants
were not told what the meals were in case preconceptions about the
nutrient content of an individual meal influenced their judgements
(Meal 1 was a frozen moussaka, Meal 2 a beef hot pot and Meal 3 a pork
pie).

Figure 7 (on p. 31) shows the levels for each nutrient in the meals, as
defined by a banding scheme devised by the Coronary Prevention
Group* (see Appendix 3, on p. 58). These bandings were not shown to
the participants.

The particular meals chosen were selected for the balance of nutrients in
them. We anticipated that people would be likely to give a high rating
for the fat and saturated fat content of Meal 1, because they are similar
on a g/100g basis to the carbohydrate and protein levels. Also we
anticipated that they would give sodium and dietary fibre low ratings
because their levels looked low compared to the other nutrients. In
contrast, we anticipated that in Meal 2, fat and saturated fats would
receive lower ratings, because they appeared to have lower levels than
protein and carbohydrate. We anticipated that sodium would receive a
low rating. For Meal 3 we anticipated that fat and saturated fat would
appear high in relation to protein. We thought that saturated fat might
receive lower ratings than fat because it looks low in relation to
carbohydrate levels. Again, we anticipated that sodium would receive a
low rating.

Questionnaire responses

The mean scores for the QI and QS sub-groups were 7.6 and 7.2
respectively, out of a possible score of 15 (there was no significant
difference between the scores for the two groups). Table 8 (on p. 32)
summarises the number of participants giving correct responses for
each nutrient in each meal.

---

* Coronary Prevention Group (1990). *Nutrition Banding*. London: The Coronary
Prevention Group.

Meal 1

	per 100g	per serving (330g)	
Energy	659kJ/157kcal	2175kJ/518kcal	
Protein	7.8g	25.7g	M
Carbohydrate	10.2g	33.7g	L
of which sugars	1.1g	3.6g	L
Fat	10.0g	33.0g	H
of which saturated fat	4.7g	15.5g	H
Dietary fibre	3.0g	9.9g	H
Sodium	0.3g	1.1g	H

Meal 2

	per 100g	per serving (330g)	
Energy	481kJ/115kcal	1587kJ/380kcal	
Protein	7.1g	23.4g	H
Carbohydrate	8.6g	28.4g	L
of which sugars	0.1g	0.3g	L
Fat	5.9g	19.5g	H
of which saturated fats	2.0g	6.6g	H
Dietary fibre	0.0g	0.0g	L
Sodium	0.2g	0.7g	H

Meal 3

	per 100g	per serving (140g)	
Energy	1487kJ/355kcal	2081kJ/497kcal	
Protein	9.4g	13.2g	M
Carbohydrate	22.8g	31.9g	L
of which sugars	0.5g	0.7g	L
Fat	26.0g	36.4g	H
of which saturated fats	11.0g	15.4g	H
Dietary fibre	0.0g	0.0g	L
Sodium	0.7g	1.0g	H

Figure 7: *Nutrition information used in task where participants made judgements about the levels of nutrients (high, medium, low) in single foods. The letters H (high), M (medium), L (low) indicate the Coronary Prevention Group bandings.*

Table 8: *Summary of number of correct judgements of nutrient levels for each meal used in single meal task (maximum 29)*
(As measured against the Coronary Prevention Group banding scheme)

			Correct responses		
	sugars	fat	saturated fats	sodium	fibre
Meal 1	15	19	12	2	5
Meal 2	20	12	4	0	n/a
Meal 3	18	24	20	9	n/a

As we anticipated, people gave fat and saturated fat 'high' ratings when their level looked high in relation to other nutrients (in Meal 1 and Meal 3), but not when they looked low in relation to the levels of other nutrients (as in Meal 2). Saturated fats received fewer 'high' responses, presumably because they always looked lower than fat.

Sodium levels were rarely judged as high, even though in all three meals the sodium levels were high according to the Coronary Prevention Group banding scheme. The increased number of 'high' responses for sodium in Meal 3 may reflect learning over the three examples.

In Meals 2 and 3 there was no dietary fibre listed, and so participants' responses of 'low' were bound to be correct. In Meal 1 the level of fibre was high according to the Coronary Prevention Group banding scheme, but not considered to be high by the participants.

It appears that most participants were trying to apply the same scale to the different nutrients in order to judge their levels. In the case of fat, saturated fats, sodium and fibre, this led to incorrect responses, when marked against the Coronary Prevention Group banding scheme. The exception was sugar, which had a low level in all these savoury meals, and fitted with the assumed scale participants were working with. So it had a consistently high level of correct responses.

8.2 Comparing two foods

The task

We presented participants with the nutrition information for two pairs of meals and asked them to judge which one of each pair would be 'a healthier choice', and which one would be the more fattening. The two comparisons are shown in Figure 8, on p. 33.

Participants were not told what the meals were. For each comparison, participants could respond with any of the following responses: A, B, 'they're the same', 'it's impossible to tell'. They were asked to explain their response.

We thought that Comparison 1 (between steak and kidney pie, in A, and a vegetable au gratin, in B) would be a relatively easy comparison to make, since A was higher than B in calories, protein, carbohydrate, fat, saturates and sodium – and lower in dietary fibre. We thought that the Comparison 2 (between potato dauphinoise, in A, and chicken wings in hot and sour sauce, in B) would be relatively difficult because both meals had the same number of calories, but while A was lower in sugars and higher in fibre than B, it was also higher in overall fat, and saturated fats, and lower in protein than B.

Comparison 1

Meal A

	per 100g
Energy	1170kJ/272kcal
Protein	9.3g
Carbohydrate	21.0g
of which sugars	0.0g
Fat	17.8g
of which saturated fat	6.9g
Dietary fibre	0.0g
Sodium	0.4g

Meal B

	per 100g
Energy	297kJ/71kcal
Protein	3.2g
Carbohydrate	8.9g
of which sugars	1.3g
Fat	2.7g
of which saturated fat	0.8g
Dietary fibre	1.4g
Sodium	0.3g

Comparison 2

Meal A

	per 100g
Energy	603kJ/142kcal
Protein	2.3g
Carbohydrate	11.8g
of which sugars	3.0g
Fat	9.6g
of which saturated fat	6.0g
Dietary fibre	0.5g
Sodium	0.1g

Meal B

	per 100g
Energy	603kJ/142kcal
Protein	14.2g
Carbohydrate	18.5g
of which sugars	9.0g
Fat	2.0g
of which saturated fat	0.7g
Dietary fibre	0.1g
Sodium	0.1g

Figure 8: Nutrition information for pairs of foods used in task to assess people's comparison across different foods.

Questionnaire responses

Participants' responses are summarised in Table 9, on p. 34.

a. Comparison 1

In Comparison 1 most participants (24) answered, as had been expected, that B was the healthier choice; and 16 of these participants gave explanations of their choice that indicated that their answer was based on differences in fat, saturated fat, sodium and fibre content. Of the 24 participants choosing B, 13 were in the QI sub-group (100% of the sub-group) and 11 in the QS sub-group (69% of the sub-group).

Twenty-six of the participants thought that A was the more fattening choice (12 in the QI sub-group and 14 in the QS sub-group – 92% and 88% of each sub-group respectively). Of the participants citing A, only nine indicated that A contained more calories than B (six from the QI sub-group and three from the QS sub-group – 46% and 19% of each sub-group respectively). Most other explanations mentioned the fat levels in the two meals; some mentioned carbohydrate, protein and dietary fibre levels.

Table 9: *Comparisons of two foods – summaries for the QI and QS sub-groups*

The numbers of participants in each group giving each response are shown on the left of each column, and the percentages on the right

| | **Comparison 1** | | | | | | | |
| | *Which would be a healthier choice?* | | | | *Which would be more fattening?* | | | |
sub-group	QI		QS		QI		QS	
Response								
A	0	(0%)	2	(13%)	12	(92%)	13	(81%)
B	11	(85%)	11	(69%)	1	(8%)	1	(6%)
Same	0	(0%)	0	(0%)	0	(0%)	0	(0%)
Impossible to tell	0	(0%)	2	(13%)	0	(0%)	2	(13%)

| | **Comparison 2** | | | | | | | |
| | *Which would be a healthier choice?* | | | | *Which would be more fattening?* | | | |
sub-group	QI		QS		QI		QS	
Response								
A	2	(15%)	1	(6%)	5	(38%)	6	(38%)
B	7	(54%)	11	(69%)	0	(0%)	2	(13%)
Same	3	(23%)	1	(6%)	6	(46%)	3	(19%)
Impossible to tell	1	(1%)	2	(13%)	0	(0%)	3	(19%)

b. Comparison 2

In Comparison 2 we expected most participants to say that it was impossible to tell whether either meal was healthier than the other and to give an explanation saying that although B was lower in fat than A, it was higher in sugar. However, a majority of participants (18) responded that B was the healthier choice, citing its lower fat level. This may have been a consequence of the emphasis on fat in the questionnaire and discussions. Indeed, some of these respondents qualified their explanations with statements showing that although they had based their judgement on fat content, they were aware that B was also high in sugar. Seven of the participants selecting B were in the QI sub-group and eleven in the QS sub-group (54% and 69% of these sub-groups respectively).

Three participants thought that A was a healthier choice (two from the QI sub-group and one from the QS sub-group). Four thought the meals were the same (three from the QI sub-group and one from the QS sub-group) and three thought it was impossible to tell (one from the QI sub-group and two from the QS sub-group). Unfortunately, participants who answered that the meals were the same or that it was impossible to tell did not explain their answers.

We expected participants to respond to the question asking which meal was more fattening by saying that both meals were the same. But only nine gave this answer. Six of these were in the QI sub-group and three in the QS sub-group (46% and 19% of these sub-groups respectively). Of the participants responding in this way five gave explanations that stated the meals were equally fattening because they had the same number of kilocalories (three in the QI sub-group and two in the QS sub-group). Of the 17 other responses (three participants made no response to this question) the majority (11) responded that A was the more fattening – four in the QI sub-group and seven in the QS sub-group (30% and 44% of each sub-group respectively). And nine of the explanations for these responses cited the amount of fat in A to account for the choice. Three participants said B was more fattening (one in the QI sub-group, two in the QS sub-group) and three (all in the QS sub-group) that it was impossible to tell.

Overall the comparison tasks suggest that people can draw on their common sense understanding of nutrition to make relatively simple judgements about differences between meals, as in Comparison 1. But this understanding is not stable, and does not provide good support for the more complex decision-making of the kind required in Comparison 2. In this case, many participants' strategy, perhaps not surprisingly given the content of the discussion groups, was to use the single factor of fat content as a yardstick for their judgements. Throughout the comparison tasks the performance of the QI sub-group was slightly better than the QS sub-group, suggesting less of a tendency to focus on single factors, and a clearer understanding of fat as one of a cluster of nutrients contributing to the nutritional and calorific value of foods.

8.3 Group discussion

Judging single foods

The discussion of this task revealed that people may encounter a number of different difficulties when trying to understand nutrition labelling.

a. Perceptual difficulties

Older participants (one of whom had particularly poor vision) found it difficult to scan the different rows of nutrition information to find the information they needed (the information was shown at the same size as it is shown in Figures 7 and 8). Some other (younger) participants said they were confused by having fat and saturated fats listed next to one another, although, generally, this was not thought to be a problem.

b. Difficulties manipulating numerical information

Some participants said they had difficulty with decimals and would have preferred the information to be presented in ounces 'without points'. The participants who complained about metric measurements were mainly, but not exclusively, older.

c. Insufficient background knowledge

Participants in all groups said they had no idea what might be high, medium or low levels for nutrients. Some participants were not clear whether there might be different scales for each nutrient, although when prompted most agreed that there were likely to be lower levels for sodium. Participants felt that they could make guesses of appropriate levels for protein, carbohydrate and fat, but that fibre and sodium would be impossible to estimate. Participants in the RS group said they ignored the sodium figures because they were so small (although contrary to these statements, the improved ability to judge sodium in the third meal compared to the first and second meals suggests that they were taking account of it).

d. Additional comments

Despite the difficulties expressed in all groups, participants were able to see that some cases were easier to judge than others, and said that they found Meal 3 easier to assess than Meals 1 and 2, a perception that is borne out by the questionnaire responses.

Participants were unsure whether or not nutrient levels were information they really needed to know. One participant in the RI group said she didn't have enough scientific understanding to be able to use nutrient levels, even if they were supplied. Some thought they might use nutrient levels if, for example, they knew they had heart trouble. One participant asked where she might get help in understanding nutrient levels, wondering if she might be told about them at Weight Watchers or slimming classes. Some thought information about nutrient levels should be available from doctors; others thought that doctors were too busy.

Comparing two foods

Participants' comments suggested that, as had been anticipated, they had found Comparison 2 more difficult than Comparison 1. They talked of the difficulty of balancing out considerations of fat and sugar.

When asked if they ever made comparisons of this kind, they said that they did not usually have the time for such detailed attention to the contents of the food they bought and that they were usually guided more by their own knowledge of 'what to steer clear of' (i.e. the strategies for healthy eating suggested in their discussion of nutrition terms) than by the information on packaging. Some said they might look for detailed information when they were buying something they had not bought before, when they would check for calories (for themselves) or additives (for their children).

Participants' conclusions about nutrition labelling

When asked if any improvements to the information on packaging might make them more likely to look at it, some participants thought that too much information was given already, particularly as they did not have sufficient background information to interpret it properly. Participants in both the RS and RI groups said it would be useful to have recommended daily allowances on packaging to help with interpreting the nutrition information. Other participants (in the RS and MS_1 groups) thought it would be useful to have that kind of information posted on a wall chart at home, where there would be time to look at it.

Participants in the MI group felt it would be far simpler to have direct warnings on foods, like government health warnings on cigarettes, as well as endorsements (such as the ones they thought were currently given by the British Heart Foundation). When asked if they would respond to such a warning, they said they thought they would, just as they were currently influenced by calorie content.

9 Assessing alternative formats for presenting nutrition information

This section of the discussion groups differed from the other sections in that participants discussed the different labelling formats first. Then after the discussion each participant completed a section of the questionnaire on the formats. It was necessary to proceed in this order so that the formats could be explained to the participants and so that they would have time to think about them before responding to the questionnaire. Accordingly the group discussion is reported before the questionnaire responses below.

9.1 Group discussion

Task

We presented the participants with a range of different graphic formats to be used in conjunction with numeric information. The formats are shown in Figure 9. Each format was shown on a single A4 page of an eight-page booklet. The format was displayed alongside a standard numeric listing (*a* in Figure 9) to represent the nutrient content of a single food, and then used to display the two comparisons used in the task to examine peoples' understanding of nutrient levels, see sub-section 8.2.

The different graphic formats were all designed to be printed in a single colour so that they could be used across a range of packaging. They were designed to be suitable for display on cardboard cartons, paper labels wrapped around tins or jars, printed tins and plastic cartons, although it was clear that many could not be applied to very small containers, such as packets of stock cubes. The formats are described below.

a. *Numeric listing* of the eight nutrients specified in the EC Nutrition Labelling Rules Directive. Horizontal rules were used to separate the fields; to emphasise the use of two units for energy content; and to emphasise the inclusion relationships between carbohydrate and sugars, and between fat and saturated fats.

b. *Bar charts* that mapped directly the quantities shown in the numeric listing for a serving. Note that the horizontal scale extended only far enough to accommodate the longest bar of the chart. It might have been better if a single scale (extending, say, up to 100g) had been used for all the examples, but this would have produced extremely wide charts using the scale shown here. The scale of the bar chart could not be compressed any further because small quantities (particularly the amount of sodium and fibre) would not have been visible on the scale. This problem has not been reported in previous attempts to examine the use of bar charts to represent nutrition information,* because sodium and fibre content have not been included, and so there have not been such different scales.

---

* See, for example, Consumers' Association, Ministry of Agriculture Fisheries and Food, National Consumer Council (1985). *Consumer attitudes to and understanding of nutrition labelling*. London: Consumers' Association.

a *Numeric listing*

	per 100 grams	per serving (330g)	
Energy	659	2175	kJ
	157	518	kcal
Protein	7.8	25.7	g
Carbohydrate	10.2	33.7	g
of which sugars	1.1	3.6	g
Fat	10.0	33.0	g
of which saturated fats	4.7	15.5	g
Dietary fibre	3.0	9.9	g
Sodium	0.32	1.1	g

b *Bar chart*

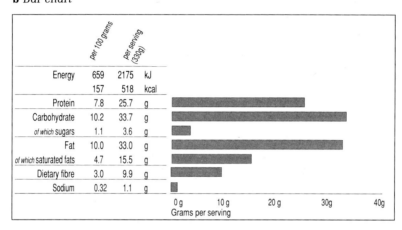

c *Banding*

	per 100 grams	per serving (330g)		
Energy	659	2175	kJ	
	157	518	kcal	
Protein	7.8	25.7	g	MEDIUM
Carbohydrate	10.2	33.7	g	LOW
of which sugars	1.1	3.6	g	LOW
Fat	10.0	33.0	g	HIGH
of which saturated fats	4.7	15.5	g	HIGH
Dietary fibre	3.0	9.9	g	HIGH
Sodium	0.32	1.1	g	HIGH

d *Banding with bars*

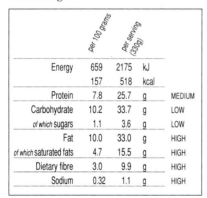

Figure 9: *Alternative formats for nutrition information*

e *Listing of RDA*

	per 100 grams	per serving (330g)		Recommended daily allowance
Energy	659	2175	kJ	
	157	518	kcal	
Protein	7.8	25.7	g	75g
Carbohydrate	10.2	33.7	g	345g
of which sugars	1.1	3.6	g	72g
Fat	10.0	33.0	g	80g
of which saturated fats	4.7	15.5	g	27g
Dietary fibre	3.0	9.9	g	30g
Sodium	0.32	1.1	g	2g

f *Percentage of RDA*

	per 100 grams	per serving (330g)		Percentage of recommended daily allowance (per serving))
Energy	659	2175	kJ	
	157	518	kcal	
Protein	7.8	25.7	g	34%
Carbohydrate	10.2	33.7	g	10%
of which sugars	1.1	3.6	g	5%
Fat	10.0	33.0	g	41%
of which saturated fats	4.7	15.5	g	57%
Dietary fibre	3.0	9.9	g	33%
Sodium	0.32	1.1	g	55%

g *Percentage RDA with bars*

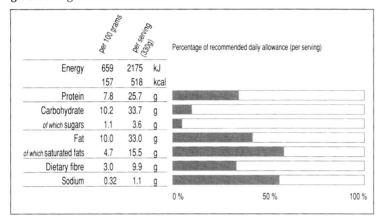

h *'Plus points'*

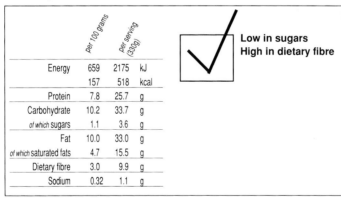

i *Selective banding*

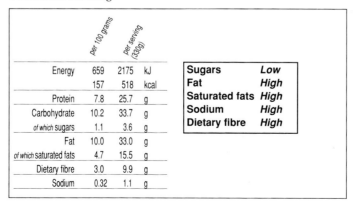

Figure 9 *(continued): Alternative formats for nutrition information*

c. *Banding* giving one word (low, medium, high) beside the numeric information for each nutrient.* A three band system was used here, and in the other variants using banding which we examined (*d* and *i*), but there is no reason why a four-band system could not have been used either here or in the other examples.

d. *Banding with bars* where the banding was emphasised by horizontal bars that extended a short, medium or longer distance from the numeric information.

e. *Listing of recommended daily allowances (RDAs)* giving the RDA (in grams) beside each nutrient.†

f. *Percentage of recommended daily allowances (RDAs)* in which the percentage of the RDA of each nutrient provided by a serving was listed beside each nutrient.

g. *Percentage of recommended daily allowances (RDAs) represented as a bar chart.* The full RDA for each nutrient was represented as an unshaded bar extending from 0% to 100%. The percentage of each RDA provided by a serving was shown as a shaded portion of the appropriate bar. Note that using a percentage representation such as this one overcomes the problems of the scale described for the bar charts in *b*.

h. *'Plus points.'* This departed from any attempt to represent all the numeric information and gave a banding for only those aspects of nutrition content that might be considered to be of positive value for health (it could include low in sugars, low in fat, low in saturated fat, low in sodium and high in fibre). Note that this listing focussed attention on one extreme of the spectrum, excluding information about content with a medium banding, and, of course, information about content that might be considered to have a negative value for health. The plus points were listed next to a large tick adjacent to the numeric listing.

i. *Selective banding.* An independent listing of bandings for nutrition content considered to have the most influence on health (sugars, fat, saturated fat, sodium and fibre). Each nutrient could be banded as low, medium or high.

Reactions to the formats

a. Numeric listing

This format was not discussed individually, but used as a reference for the discussion of other formats.

b. Bar charts

Most participants initially thought the bar charts were an improvement on the numeric listing, commenting that the charts allowed them to see 'at a glance' what nutrients were high or low, and so would be helpful when they were in a hurry and for people with poor vision. However participants in MS2 thought the bars looked unnecessarily complicated and would only be useful for people who knew what they were looking

---

* The bandings were according to the scheme of the Coronary Prevention Group, see Appendix 3, on p. 58.

† The recommended daily allowances were derived from the dietary targets used as a basis for the banding scheme, see Appendix 3. Note that after phase 1 of the study had been carried out, the Ministry of Agriculture Fisheries and Food adopted the term *dietary reference values* as a standard term for recommended daily allowances and other similar terms, and the new term was used throughout phase 2 of the study.

for, who could quickly pick out the bars relevant to them (for example, people on restricted diets). One participant said 'ordinary people, all they'd really like to know is, is it good or bad?'

When asked if it was easier or more difficult to make judgements about single foods with bar charts, groups other than MS2 said they thought that it was easier. One participant cited how with one of the bar charts she could immediately see that saturated fats made up about half of the total fat content.

All the groups were asked about the difficulty of representing fibre and sodium (which have relatively low values as g/100g) on the same scale as nutrients such as carbohydrate and fat. Most did not see this as a real problem, saying that they would simply have to look at the numeric listing if they need information about those nutrients. However one participant realised that the usefulness of the format would depend on the user's background knowledge, and ability to interpret the meaning of bars of different length. She said 'it's only by knowing that the quantity of fibre is a normal amount in relation to other things that you can decide on it.' She went on to recommend a system that gave recommended daily intake as a guide to users. Another participant suggested a colour coded bar system where nutrients with quantities that were considered not to be beneficial to health were shown in a specific colour, say, red. However she conceded that manufacturers might not be in favour of such a system.

One participant, who was a diabetic, thought the bars helped her to see quickly whether particular nutrients were present or not in single foods. This was important to her since she needed to know whether or not foods contained sugar, but could be important for other groups of people who needed quick recognition of the presence or absence of individual nutrients and some rough indication of quantity.

When asked whether it was easier or more difficult to make comparisons with bar charts than with numeric listings, most groups thought it was easier. However when they were encouraged to try and make a simple comparison during the discussion (between the amount of carbohydrate in two meals) they found that it was difficult to scan across two displays. Participants in the RS group made the comparison incorrectly, and were doubtful about using bands for comparisons. The MI group thought the problems in scanning across two displays resulted from not being familiar with the order in which the nutrients were listed, so they had to keep checking by scanning within each display to make sure that the bar they were looking at tallied with the nutrient they were comparing. It would have been easier, they said, if the nutrient names had been right next to the bars. This informal test of the bars and the ensuing comments belie the initial impression that bar charts might be particularly useful for people in a hurry and people with poor vision.

c. Banding

Having discovered the difficulties inherent in the bar charts most (but not all) participants thought that banding was a substantial improvement, both on the simple numeric listing and on the numeric listing with bar chart. One participant who had poor vision found it difficult to see the words, although she still found the format an improvement, since she could judge what the words were by their length.

When asked how easy or difficult banding made judgements about single foods, participants in MS1 said that they would still need to be sure that, for example, high fibre was desirable, high fat undesirable, and so on. Other groups did not think these different mappings of nutrient level on to desirability for health were likely to be a problem. One participant pointed out that there might be a confusion since fibre (where a high level was desirable) was listed between saturated fat and sodium (where a low level was desirable). The MI group thought that although most people knew whether high or low levels of the different nutrients were desirable, it might still be better to have an evaluation of the level 'good, bad, indifferent' rather than simply stating the level.

The MI group asked how the banding system had been worked out. When the method of banding was briefly explained they said they felt that the system would need to be endorsed by some authority so that people trusted it.

When asked whether comparisons were easier or more difficult with banding than with the other types of information they had seen, all groups said they thought they were easier.

d. Banding with bars

The participants were divided about whether this was any improvement on the simple, verbal banding system. Those who thought it was an improvement felt that the bars had a visual impact that was missing from the verbal banding (especially older participants who had found the words alone difficult to see). One participant commented that if the labelling was poorly printed the bars would stand out even if the letters were not clear. Other participants could not see how this format would make decisions about single foods or comparisons easier. Participants in the MS_2 group and the MI group thought the bars were more difficult to read than the verbal banding (possibly because the banding words did not appear in a single column). The MI group asked whether the banding information needed to be next to the numerical listing, suggesting that vertical bars might be easier to use than horizontal ones if they were displayed in a separate listing.

e. Listing of recommended daily allowances

In many of the groups, individual participants had already suggested spontaneously that it would be useful to have the RDA for nutrients displayed within the nutrition labelling. However when the groups were presented with such as display many thought it was too complicated.

Some people objected to having yet another list of numbers (especially a listing in grams) to negotiate and said they would not have time to look at this amount of information in the supermarket, although they might find it useful at home. Others objected to the complexity of the decision-making the display required: one commented 'this is one food, and you've got to then compare that with the rest of the day, with food that may not have had information on, so you've got to guess how much your cornflakes came to so you know how near your RDA you are...'. Others thought it might make people anxious, rather than help them: 'it might make you think you've got to make up an allowance, for example there's only 40g of carbohydrate and the RDA is 345g. It might make you worry.' Some participants asked whether it was possible to give RDAs since different people (such as manual and office workers) had different needs.

Despite these objections some people were still in favour of the principle of showing RDAs for people who wanted detailed information. Two suggestions were made for systems that graphically represented servings as proportions of the RDA: 'it would be useful to have colour coding to show how much you should have in a day (say in blue) with the amount of food on top of it (say in red); 'what about having a point or star system instead of these numbers...say ten points are recommended and the packet says how many (points) this product contains.'

f. Percentages of recommended daily allowances

Some participants could see the advantage of this format over the listing of RDAs in grams. There was still a feeling that this type of information would be more useful for consultation at home than in shopping, especially as the RDA might need to be adapted for different family members' needs. Other participants thought that percentages were too complicated. One participant suggested that fractions might be easier to understand, but changed her mind when asked about the complexity of the fractions that she would be prepared to deal with. As with the listing of RDAs, many participants were put off by the amount of information given: one commented 'you'd be paranoid by the time tea time came. That's too blatant that is.'

g. Percentages of recommended daily allowances represented as bar charts

Most participants found this representation difficult to understand and were discouraged because it looked technical. This was in spite of the fact that it resembled the proportionally-based representations of RDAs suggested by the RS group when they were discussing listings of RDAs. The MI group, who had found the numerical or percentage representations of RDAs easier to handle than most other groups, suggested that this might be a useful format for people who found percentages difficult to understand. But its reception by groups who had found percentages complicated suggested that it was not. Conversely, participants who had found percentages difficult thought that this bar chart would be helpful to people who really needed to watch what they ate. However, the participant in the MI group who had diabetes pointed out that although the format looked technical, the scale used was too small, and not clearly enough marked, to be of genuine assistance to her.

h. 'Plus points'

This format appealed to most groups initially: it looked direct, simple, and non-numerical compared to the formats they had examined previously. However when participants began to understand that it only represented a limited amount of nutrition information they were less convinced. It took some time to explain its principles to some groups (RS and MS_2).

The use of a single tick to draw attention to the plus points was not liked: one participant commented, 'if they've got the same tick, they look the same, even though they've got different content.' Some people thought the tick might make them pick up a tin or packet, after which they would look for more detailed information. But others felt that they would cease to notice the tick if it was used on all foods, and others felt it 'looked a bit tacky.'

All groups thought that it would be useful to complement the tick with a minus sign or cross listing 'minus points' in the style of a government health warning. This was in spite of their objection to percentages of RDAs as being off-putting and anxiety-provoking. The MI group suggested using stars or colours to cue instant recognition of foods that were, for example, low in salt or fat (they described systems they thought were used by some supermarkets for marking low salt foods with coloured bands). This could then be followed up by examination of the detailed nutrition labelling.

i. Selective banding

All the groups preferred this representation to the 'plus points' format – describing it as simple and bold. They were asked to compare it to the banding system used for all nutrients (c). Some felt this format was more direct and easy to use because it was separate from the numeric listing. The RS group thought that although this format was likely to be quicker to use it would be better to see all nutrients banded. The RI group and some participants in MS_1 thought that banding with bars was better than this format because it gave more complete information and also had greater visual impact.

9.2 Individual preference ratings

Task

We presented the participants with a summary page showing all the formats they had discussed. We asked participants to list which of the formats would provide the best source of information in a range of different circumstances – shopping, decision-making at home, when in a hurry, when there was plenty of time, for decisions about single foods and for comparisons between foods. We also asked participants to say which of the formats were the most informative, the least informative, the easiest to use and the most difficult to use. Participants could respond with more than one format to each question, and could respond with the same format or formats for more than one question.

Questionnaire responses

Participants' responses for each question are summarised in Table 10, on p. 45. The preferred format for most of the decision-making tasks was the selective banding format (shown in *i*). It was usually followed by banding with bars (shown in *d*) or verbal banding (shown in *c*). There was a different pattern of preferences, however, when participants were asked about information to support decisions 'when there is plenty of time'. Here, banding with bars (*d*) was the first preference, followed by the listing of percentages of RDAs (*f*) as a second preference, and a joint third preference of a listing of RDAs in grams (*e*) or bars showing percentages of RDAs (*g*). Note also that the listing of percentage of RDAs (*f*) is joint second preference for decisions at home (presumably also because there would be more time for consultation than in shopping). And a very different kind of representation, 'plus points' (*h*) is joint third preference for decisions about single foods.

Table 10: *Rank ordering of participants' preferences for formats*

Key to formats, as they are shown in Figure 9:
num. listing – numeric listing (a), bar chart (b), banding (c), banding/bars – banding with bars (d), list RDA – parallel listing of RDAs in grams (e), % of RDA – parallel listing of percentages of RDAs (f), bar % RDA – bar representation of percentages of RDAs (g), plus points (h), sel. banding – selective banding (i).

The numbers in parentheses show the number of participants selecting each format (maximum 28).

for shopping		*for decisions at home*		*when in a hurry*		*when there is plenty of time*		*for decisions on single foods*		*for comparisons between foods*	
sel. banding	(13)	sel. banding	(12)	sel. banding	(12)	banding/bars	(11)	sel. banding	(10)	sel. banding	(14)
bands/bars	(11)	banding	(7)	banding/bars	(10)	% of RDA	(8)	banding	(8)	banding/bars	(9)
banding	(8)	banding/bars	(7)	banding	(6)	list RDA	(6)	banding/bars	(6)	banding	(7)
% of RDA	(5)	% of RDA	(7)	plus points	(5)	bar % RDA	(6)	plus points	(6)	% of RDA	(5)
num. listing	(3)	num. listing	(4)	bar chart	(3)	banding	(5)	list RDA	(5)	plus points	(4)
list RDA	(3)	list RDA	(3)	% of RDA	(3)	sel. banding	(4)	% of RDA	(5)	bar chart	(3)
plus points	(3)	plus points	(2)	list RDA	(2)	plus points	(2)	bar % RDA	(5)	list RDA	(2)
bar chart	(2)	bar chart	(1)	bar % RDA	(1)	num. listing	(1)	num. listing	(1)	num. listing	(1)
bar % RDA	(1)	bar % RDA	(1)					bar chart	(1)	bar % RDA	(1)

most informative		*least informative*		*easiest to use*		*most difficult to use*	
banding/bars	(10)	num. listing	(15)	banding/bars	(13)	num. listing	(10)
% of RDA	(8)	plus points	(11)	sel. banding	(9)	list RDA	(9)
sel. banding	(8)	bar chart	(4)	banding	(6)	% of RDA	(9)
banding	(6)	list RDA	(3)	plus points	(4)	bar chart	(7)
list RDA	(4)	% of RDA	(3)	bar chart	(3)	bar % RDA	(6)
bar % RDA	(4)	bar % RDA	(3)	% of RDA	(3)	banding	(3)
bar chart	(1)	sel. banding	(2)	list RDA	(2)	banding/bars	(2)
plus points	(1)	banding	(1)	bar % RDA	(2)	plus points	(2)
		banding/bars	(1)	num. listing	(1)	sel. banding	(1)

Table 11 shows the responses for the QI and QS sub-groups. Although these follow similar trends, the QS sub-group showed a more consistent preference for the selective banding format (*i*). The listing of percentages of RDAs (*f*) always appeared higher in the QI sub-group's list of preferences than in the QS sub-group's list.

Participants' preferences for banding with bars (*d*) and selective banding (*i*) were evident in their overall evaluations of how informative and easy to use the different formats were (Table 10). The listing of percentages of RDAs (*f*) was considered to be an informative format but was also considered to be relatively difficult to use. The separate QI and QS sub-group results (Table 11) show, again, that this format is favoured more by the QI sub-group than the QS sub-group. Note also that the bar representation of percentages of RDAs (*g*) is rated as the most difficult to use by the QI sub-group, even though it is a direct graphic translation of their favoured listing of percentages of RDAs (*f*).

Table 11: *Rank ordering of QI (questionnaire interested) and QS (questionnaire standard) sub-groups' preferences for formats*

Key to formats as they are shown in Figure 9.
num. listing – numeric listing (a), bar chart (b), banding (c), banding/bars – banding with bars (d), list RDA – parallel listing of RDAs in grams (e), % of RDA – parallel listing of percentages of RDAs (f), bar % RDA – bar representation of percentages of RDAs (g), plus points (h), sel. banding – selective banding (i).

The percentages in each sub-group selecting each format are shown (QI n = 12; QS n = 16).

for shopping				for decisions at home			
QI sub-group		QS sub-group		QI sub-group		QS sub-group	
banding	42%	sel. banding	50%	banding/bars	42%	sel. banding	56%
bands/bars	42%	banding/bars	38%	% of RDA	42%	banding	25%
sel. banding	42%	banding	19%	banding	25%	num. listing	13%
% of RDA	25%	num. listing	13%	sel. banding	25%	banding/bars	13%
list RDA	17%	% of RDA	13%	num. listing	17%	% of RDA	13%
num. listing	8%	list RDA	6%	list RDA	17%	list RDA	6%
bar % RDA	8%	plus points	6%	bar % RDA	8%	plus points	6%
plus points	8%			plus points	8%		

when in a hurry				when there is plenty of time			
QI sub-group		QS sub-group		QI sub-group		QS sub-group	
sel. banding	50%	banding/bars	48%	bands/bars	42%	banding/bars	38%
banding	33%	sel. banding	38%	% of RDA	33%	list RDA	25%
% of RDA	25%	plus points	19%	banding	17%	% of RDA	25%
banding/bars	25%	banding	13%	list RDA	17%	bar % RDA	25%
plus points	17%	list RDA	6%	bar % RDA	17%	banding	19%
list RDA	8%			sel. banding	17%	sel. banding	13%
bar % RDA	8%			num. listing	8%	plus points	6%
				plus points	8%		

Table 11 *(continued): Rank ordering of QI (questionnaire interested) and QS (questionnaire standard) sub-groups' preferences for formats*

Key to formats as they are shown in Figure 9.
num. listing – numeric listing (a), bar chart (b), banding (c), banding/bars – banding with bars (d), list RDA – parallel listing of RDAs in grams (e), % of RDA – parallel listing of percentages of RDAs (f), bar % RDA – bar representation of percentages of RDAs (g), plus points (h), sel. banding – selective banding (i).

The percentages in each sub-group selecting each format are shown (QI n = 12; QS n = 16).

for decisions on single foods		for comparisons between foods	
QI sub-group	QS sub-group	QI sub-group	QS sub-group
banding/bars 50%	sel. banding 31%	sel. banding 50%	sel. banding 50%
sel. banding 42%	banding 25%	banding 42%	banding/bars 38%
banding 33%	banding/bars 25%	% of RDA 33%	banding 13%
% of RDA 33%	plus points 19%	banding/bars 25%	bars 13%
list RDA 25%	list RDA 13%	list RDA 17%	plus points 13%
plus points 25%	% of RDA 6%	plus points 17%	% of RDA 6%
num. listing 8%		bar chart 8%	
bar % RDA 8%		bar % RDA 8%	

most informative		least informative	
QA sub-group	QS sub-group	QA sub-group	QS sub-group
% of RDA 42%	banding/bars 38%	num. listing 58%	num. listing 50%
bands/bars 33%	sel. banding 25%	plus points 42%	plus points 25%
sel. banding 33%	banding 19%	bar chart 25%	banding/bars 13%
banding 25%	% of RDA 19%	bar % RDA 25%	bar chart 6%
list RDA 17%	bar % RDA 19%	list RDA 17%	list RDA 6%
num. listing 8%	list RDA 13%	banding/bars 8%	% of RDA 6%
bar % RDA 8%			
plus points 8%			

easiest to use		most difficult to use	
QA sub-group	QS sub-group	QA sub-group	QS sub-group
sel. banding 58%	banding/bars 50%	bar % RDA 42%	num. listing 44%
banding 42%	bar chart 13%	bar chart 33%	list RDA 38%
bands/bars 42%	plus points 13%	num. listing 25%	% of RDA 38%
% of RDA 25%	sel. banding 13%	list RDA 25%	bar chart 19%
plus points 17%	banding 6%	% of RDA 25%	banding 13%
bar chart 8%	list RDA 6%	banding/bars 17%	bar % RDA 6%
list RDA 8%	bar % RDA 6%	plus points 17%	
bar % RDA 8%	num. listing 6%	banding 8%	
		sel. banding 8%	

Appendix 1

Recruitment questionnaire for discussion groups in phase 1

1. Do you do most of the food shopping for you and your family?

2. Do you (also) make the decisions about what to buy, or is a list made by other members of the family?

3. Are you aware of any link between the food you eat and the chance of developing coronary heart disease?

4. Can you tell me the main things you need to do to eat for a healthy heart?

    Check for:

    A   change to reduced fat milk

    B   substitute polyunsaturated margarine for butter or other animal fats

    C   reduce fatty red meats, substitute chicken/fish

    D   reduce overall consumption of full fat dairy products

    E   reduce salt (sodium) consumption

5. Do you or any member of your family follow a special diet?

6. Have you, or anyone else in your household, had your cholesterol level measured?

7. How many people are there in your household? Who are they?

8. What is your occupation? And the occupation of your husband/wife/partner?

9. Availability to attend discussion groups and details of name, address, phone number.

Appendix 2

Discussion group questionnaire

Some details about you

- Your sex: male   female   (*please circle as appropriate*)
- Your age: 18-25   26-35   36-45   46-55   56-65   (*please circle as appropriate*)
- At what age did you leave school? _____
- What qualifications, if any, have you gained since you left school?

  _____

- Do you follow any special diet?   yes   no   (*please circle as appropriate*)
- If you follow a special diet, please circle to show if it is

  vegetarian   slimming   religious   low-fat

  other _____ (*please state what kind of diet*)
- How often do you weigh foods (in cooking)?

  at least once a week   at least once a month   at least once a year   never

  (*please circle as appropriate*)
- Rate on a scale of 1 (low) - 7 (high) your interest in cooking

  (*please circle the number that represents your level of interest*)

1	2	3	4	5	6	7

  I am not interested
  in cooking

  I am very interested
  in cooking

- Rate on a scale of 1 (low) - 7 (high) your interest in healthy eating
  (*please circle the number that represents your level of interest*)

1	2	3	4	5	6	7

  I am not interested
  in healthy eating

  I am very interested
  in healthy eating

Below is the nutrition information from the package for a prepared meal

	per 100g
Energy	659kJ/157kcal
Protein	7.8g
Carbohydrate	10.2g
of which sugars	1.1g
Fat	10.0g
of which saturated fat	4.7g
Dietary fibre	3.0g
Sodium	0.3g

- Do you look at this kind of information on food packaging?

  never   occasionally   frequently   always   (*please circle as appropriate*)

- If you do look at this kind of information when do you look at it?

  when shopping

  when planning meals at home

  other _____ (*please state*)

- If you do look at this kind of information what particular things are you usually trying to find out?

  _____

  _____

  _____

Which portion weighs 100 grams?

From the portions of different sizes for each food, pick out the one that you think weighs **100 grams**, and circle the appropriate letter on the table below to show which one you have chosen.

apples	a	b	c
beef-burgers	a	b	c
gateau	a	b	c
potatoes	a	b	c
pizza	a	b	c
marmalade	a	b	c
baked beans	a	b	c
butter	a	b	c
fish in sauce	a	b	c

Average portions

For each of these foods say whether you think you would usually eat -

less than the average portion

about the same as the average portion

more than the average portion

I would normally eat

beef-burgers	less	about the same	more
gateau	less	about the same	more
pizza	less	about the same	more
marmalade	less	about the same	more
baked beans	less	about the same	more
fish in sauce	less	about the same	more

Nutrition terminology

Tick in the appropriate column to show how well you understand the meaning of the following nutrition terms:

	I fully understand	I roughly understand	I do not understand
energy			
kilojoules			
kilocalories			
protein			
carbohydrate			
lectin			
sugars			
fat			
saturated fat			
legnose			
dietary fibre			
sodium			

When you make a decision to buy or use a particular food would information about any of the following influence your decision?

	would influence my decision	would not influence my decision
information about:		
energy		
kilojoules		
kilocalories		
protein		
carbohydrate		
sugars		
fat		
saturated fat		
dietary fibre		
sodium		

What other information about the content of food might influence your decision?

_____

_____

_____

_____

_____

Circle the appropriate responses below the following statements

Tick here to show
statements you found
difficult to decide about

It's probably good for the heart to eat more dietary fibre

True   False   It makes no difference   I don't know

It's probably good for the heart to eat more sodium

True   False   It makes no difference   I don't know

It's probably good for the heart to eat fewer saturates

True   False   It makes no difference   I don't know

It's probably good for the heart to eat more salt

True   False   It makes no difference   I don't know

It's probably good for the heart to eat food with more kilojoules

True   False   It makes no difference   I don't know

It's probably good for the heart to eat less saturated fat

True   False   It makes no difference   I don't know

It's probably good for the heart to eat more fat

True   False   It makes no difference   I don't know

It's probably good for the heart to eat more carbohydrate

True   False   It makes no difference   I don't know

It's probably good for the heart to eat food with more kilocalories

True   False   It makes no difference   I don't know

It's probably good for the heart to eat more sugars

True   False   It makes no difference   I don't know

It's probably good for the heart to eat less protein

True   False   It makes no difference   I don't know

It's probably good for the heart to eat food with more energy

True   False   It makes no difference   I don't know

The following is the nutrition information from a prepared meal

	per 100g
Energy	848 kJ/202 kcal
Protein	5.1g
Carbohydrate	10.6g
of which sugars	1.0g
Fat	15.8g
of which saturated fat	1.6g
Dietary fibre	1.0g
Sodium	0.5g

Why are the words 'of which' used in front of the word 'sugars'?

_____

_____

_____

_____

Why are the words 'of which' used in front of the words 'saturated fat'?

_____

_____

_____

_____

Please look at the following nutrition information for a prepared meal:

	per 100g	per serving (330g)
Energy	659kJ/157kcal	2175kJ/518kcal
Protein	7.8g	25.7g
Carbohydrate	10.2g	33.7g
of which sugars	1.1g	3.6g
Fat	10.0g	33.0g
of which saturated fat	4.7g	15.5g
Dietary fibre	3.0g	9.9g
Sodium	0.3g	1.1g

Tick here to show questions
you find difficult to decide on

Is this food high, medium or low in sugars?
    high   medium   low
_____

Is this food high, medium or low in fat?
    high   medium   low
_____

Is this food high, medium or low in saturated fat?
    high   medium   low
_____

Is this food high, medium or low in sodium?
    high   medium   low
_____

Is this food high, medium or low in fibre?
    high   medium   low
_____

Please look at the following nutrition information for a prepared meal:

	per 100g	per serving (330g)
Energy	481kJ/115kcal	1587kJ/380kcal
Protein	7.1g	23.4g
Carbohydrate	8.6g	28.4g
of which sugars	0.1g	0.3g
Fat	5.9g	19.5g
of which saturated fats	2.0g	6.6g
Dietary fibre	0.0g	0.0g
Sodium	0.2g	0.7g

Tick here to show questions
you find difficult to decide on

Is this food high, medium or low in sugars?
    high   medium   low

Is this food high, medium or low in fat?
    high   medium   low

Is this food high, medium or low in saturated fat?
    high   medium   low

Is this food high, medium or low in sodium?
    high   medium   low

Is this food high, medium or low in fibre?
    high   medium   low

Please look at the following nutrition information for a prepared meal:

	per 100g	per serving (140g)
Energy	1487kJ/355kcal	2081kJ/497kcal
Protein	9.4g	13.2g
Carbohydrate	22.8g	31.9g
of which sugars	0.5g	0.7g
Fat	26.0g	36.4g
of which saturated fats	11.0g	15.4g
Dietary fibre	0.0g	0.0g
Sodium	0.7g	1.0g

Tick here to show questions
you find difficult to decide on

Is this food high, medium or low in sugars?
    high   medium   low

Is this food high, medium or low in fat?
    high   medium   low

Is this food high, medium or low in saturated fats?
    high   medium   low

Is this food high, medium or low in sodium?
    high   medium   low

Is this food high, medium or low in fibre?
    high   medium   low

Here is the nutrition information for two prepared meals

Meal A	per 100g	Meal B	per 100g
Energy	1170kJ/272kcal	Energy	297kJ/71kcal
Protein	9.3g	Protein	3.2g
Carbohydrate	21.0g	Carbohydrate	8.9g
of which sugars	0.0g	of which sugars	1.3g
Fat	17.8g	Fat	2.7g
of which saturated fat	6.9g	of which saturated fat	0.8g
Dietary fibre	0.0g	Dietary fibre	1.4g
Sodium	0.4g	Sodium	0.3g

Would either meal be a healthier choice?

A　　B　　they're the same　　It's impossible to tell

Give reasons for your answer

_____

_____

_____

Is either one more fattening than the other?

A　　B　　they're the same　　It's impossible to tell

Give reasons for your answer

_____

_____

_____

Here is the nutrition information for two prepared meals

Meal A	per 100g	Meal B	per 100g
Energy	603kJ/142kcal	Energy	603kJ/142kcal
Protein	2.3g	Protein	14.2g
Carbohydrate	11.8g	Carbohydrate	18.5g
of which sugars	3.0g	of which sugars	9.0g
Fat	9.6g	Fat	2.0g
of which saturated fat	6.0g	of which saturated fat	0.7g
Dietary fibre	0.5g	Dietary fibre	0.1g
Sodium	0.1g	Sodium	0.1g

Would either meal be a healthier choice?

A　　B　　they're the same　　it's impossible to tell

Give reasons for your answer

_____

_____

_____

Is either one more fattening than the other?

A   B   they're the same   it's impossible to tell

Give reasons for your answer

_____

_____

_____

Look at the different formats of nutrition information that we have discussed.* Each has a letter by it to identify it. In the table below circle the letters to show your answers to the questions.

(You may need to circle more than one letter to answer some questions, and you may circle the same letters in more than one question).

Which format(s) would you find most useful when you are shopping?
   a   b   c   d   e   f   g   h   i

Which format(s) would you find most useful when you are deciding what to eat at home?
   a   b   c   d   e   f   g   h   i

Which format(s) would you find most useful when you are in a hurry?
   a   b   c   d   e   f   g   h   i

Which format(s) would you find most useful when you have plenty of time?
   a   b   c   d   e   f   g   h   i

Which format(s) are the most informative?
   a   b   c   d   e   f   g   h   i

Which format(s) are the least informative?
   a   b   c   d   e   f   g   h   i

Which format(s) are the easiest to use?
   a   b   c   d   e   f   g   h   i

Which format(s) are the most difficult to use?
   a   b   c   d   e   f   g   h   i

Which format(s) would be most helpful when you are making decisions about single foods?
   a   b   c   d   e   f   g   h   i

Which format(s) would be most helpful when you are comparing between foods?
   a   b   c   d   e   f   g   h   i

---

* The formats are shown in Figure 9, pp. 38-39 of this report

Appendix 3

Nutrition bandings and dietary reference values used in studies

Nutrition bandings

The bandings are those devised by the Coronary Prevention Group (1990).* Note that dietary targets have been revised (by the 1991 COMA report)† since these banding levels were devised, and banding levels may be adjusted accordingly.

Nutrient	Dietary target	Band			
		low	medium-low	medium-high	high
Protein	12.5% energy	<6.25	6.25-12.49	12.5-18.75	>18.75
Carbohydrate	57.5% energy	<28.75	28.75-57.49	57.5-86.25	>86.25
Sugars	12% energy	<6.0	6.0-11.9	12.0-18.0	>18.0
Fat	30% energy	<15.0	15.0-29.9	30.0-45.0	>45.0
Saturated fats	10% energy	<5.0	5.0-9.9	10.0-15.0	>15.0
Dietary fibre	30g/10MJ	<15.0	15.0-29.9	30.0-45.0	>45.0
Sodium	2g/10MJ	<1.0	1.0-1.9	2.0-3.0	>3.0

In the discussion groups, only three levels of banding were used: high, medium, and low. The medium band was the combined medium-high and medium-low bands shown here.

Dietary reference values (recommended daily allowances)

After phase 1 of the study had been carried out, the Ministry of Agriculture Fisheries and Food adopted the term dietary reference values as a standard term for recommended daily allowances and other similar terms. In phase 1 of the study, dietary reference values were referred to as recommended daily allowances, but the new term was used throughout phase 2.

The dietary reference values were derived by expressing the dietary target for a 2,400kcal (10MJ) per day diet in grams. Note that these values pre-date the 1991 COMA report.

	g/10MJ g/day
Protein	75
Carbohydrate	345
Sugars	72
Fat	80
Saturated fats	27
Dietary Fibre	30
Sodium	2

---

* Coronary Prevention Group (1990). *Nutrition Banding. London: The Coronary Prevention Group.*
† Committee on Medical Aspects of Food Policy (1991) *Dietary reference values for food energy and nutrients for the United Kingdom.* London: HMSO.

References

Commission of the European Communities (1990). *Council Directive of 24 September 1990 on nutrition labelling for foodstuffs*. 90/496/ EEC.

Committee on Medical Aspects of Food Policy (1991) *Dietary reference values for food energy and nutrients for the United Kingdom*. London: HMSO.

Consumers' Association, Ministry of Agriculture Fisheries and Food, National Consumer Council (1985). *Consumer attitudes to and understanding of nutrition labelling*. London: Consumers' Association.

Coronary Prevention Group (1990). *Nutrition Banding*. London: The Coronary Prevention Group.

Crawley, H. (1988). *Food portion sizes*. London: HMSO.

Phase 2: Report of experimental studies

1 Introduction

1.1 Reasons for the research

The EC Nutrition Labelling Rules Directive (1990)* regulates the presentation of nutrition information on food packaging, specifying when nutrition information must be given, the nutrients that must be shown, the order in which they should be presented, and their units. It also provides for the possibility of supplementing numeric nutrition information with information in graphical form, according to formats that are to be decided.

The aim of this, second phase of the project was to examine ways of presenting nutrition information that might help consumers make decisions about foods to use or buy. Investigations of different graphic representations formed a large part of the study. However, we did not examine graphic formats in isolation, but how they might be used in conjunction with numeric information. We also examined verbal information (banding words, such as high, medium, low) used to supplement numeric information, already used by at least one retailer to show the levels of nutrients in foods. We were aware that people prefer text to numeric, or graphic representations of information.

We tested the different information formats in tasks that resembled those people carry out when making decisions about foods: judging single foods, or comparing foods. Most other studies of nutrition information have gathered people's opinions about different formats, and how easy they might be to use, rather than examining how people use nutrition information. However there is ample evidence in studies of information design that people's opinions about the effectiveness of different formats do not predict performance. A recent study by Levy and Schucker (1991)† has shown this divergence of preference and performance specifically for nutrition information, in tasks involving the comparisons of foods.

In phase 1 of this study we gathered people's opinions about different information formats in a pre-test study, using combined discussion group and questionnaire techniques. The first phase had helped to isolate some common problems people encounter when using nutrition information, which we tried to address when preparing the different information formats tested here. The relevant findings of the previous phase and their influence on the second phase are summarised in section 1.2.

---

* Commission of the European Communities (1990). *Council Directive of 24 September 1990 on nutrition labelling for foodstuffs. 90/496/EEC.*
† Levy, A.S. and Schucker, R.E. (1991). *An experimental evaluation of nutrition label formats: performance and preference.* Presentation to National Food Processors Association: Scientific Forum, Chicago, Illinois.

1.2 Relevant findings of preparatory discussion group study

Background knowledge of nutrition issues

Most participants (from the same population as the people tested in the research described here) had a broad understanding of issues underlying nutrition and health: for example, they were likely to know that it was wise to reduce consumption of fat, saturated fats, sugar and sodium, and increase fibre intake. But they were unlikely to be able to give details of the relationship between carbohydrate and sugars, or fat and saturated fats. They felt they were unlikely to be confused by the different mappings of individual nutrients in evaluating foods: for example that high levels of fat, sugar etc. are considered bad for health, but high levels of fibre considered good. The understanding they had was adequate for them to be able to make choices, such as to avoid foods that are identifiable as being high in fat or sugar, if they wished.

Interpretation of numeric information – banding systems

While participants' general knowledge of nutrition issues helped them make some choices about foods they knew had particularly high or low levels of nutrients, this knowledge did not always help them use numeric nutrition information because they did not know what quantities (for example, grams per 100 grams) represented high or low levels of nutrients. They tended to judge all nutrients on the same scale, which resulted in their making errors, particularly for nutrients, such as saturated fats, fibre, and sodium, where relatively few grams per 100 grams or per serving constituted a high intake. So they needed additional information that interpreted numeric information. Banding systems meet this need, and we decided to examine a range of verbal and graphic banding formats in the present research.

Evaluation of numeric information – (health warning) systems

Some participants thought that representations that evaluated the levels of nutrients on health grounds would be useful (even though they were not confused about the significance for health of different mappings of nutrients). They envisaged systems in which ticks were used for nutrient levels that were considered healthy, and crosses for levels considered unhealthy. Such systems would be difficult to apply effectively (many nutrients would fall into a middle band, where neither ticks nor crosses were appropriate). Consequently, we investigated a different evaluative sytem where different banding levels were mapped on to appropriate star ratings, so that people could use them to assess foods on the basis of 'the more stars the better'.

The need for detail – dietary reference values

Some participants were suspicious of the categorical interpretations of numeric data provided by banding systems, and wanted more detailed interpretations. So we examined systems in which levels of nutrients were listed alongside dietary reference values* or shown as a percentage of the dietary reference value. These systems were more complex, but gave a fuller account of the nutrient levels of foods, than banding systems.

---

* In the course of this research the Committee on the Medical Aspects of Food Policy published their *Dietary Reference Values for Food Energy and Nutrients of the United Kingdom*. We have therefore used the term dietary reference value (DRV) in the second phase of this report, where we used recommended daily allowance (RDA) in the first.

Use of graphic representations

Participants were divided about whether graphic representations were desirable or not. Some felt they added unnecessary clutter to complex displays, others felt their 'visual impact' would speed up decision-making. Whether or not graphic representations are helpful to consumers depends on the quality of the particular representations (see Macdonald-Ross, 1977).* The details of any particular graphic representation are usually a consequence of a range of constraints, including the nature of the information to be represented, current traditions for representing different kinds of information, assumptions about how the user may expect to see information displayed, and the preferences of the people producing the formats. We recorded the decisions made in the design of the graphic formats in order to give as full a background as possible to the materials tested (see Appendix 1).

1.3 Outline of the methods used in the experimental studies

The tasks

In this second phase, we tested nutrition information presented in different formats, in short decision-making tasks. Participants were asked to judge how wise a choice individual foods might be for a healthy diet, or which of two or three foods might be the wiser choice for a healthy diet. Immediately after they had made their decision they had to give the reasons why they had made that decision. This task contrasted with the task used by Levy and Schucker (1991), whose participants were shown nutrition information for two products at a time, and asked to identify the differences between the two products, only assessing their relative healthiness as a secondary task. Levy and Schucker describe their task as an 'information search' task. Information search was a major component of our task, but the search was made in the context of decision-making that resembled real-life decision-making about the relative healthiness of foods.

The nutrition information tested was for real foods that can be bought currently in British shops. In most of the tests, participants were shown the nutrition information on flat sheets of paper. The name of the food was always given with the nutrition information. We felt it was important that people should know what foods they were judging because in real decision-making tasks people's choices are bound to be influenced by their expectations about individual foods. For example, they may expect breakfast cereals to be a source of fibre, puddings to be high in sugar, and so on. One indication that a particular format for nutrition information was failing to help consumers in their decision-making would be if participants tended to base their judgements more on their preconceptions about individual foods than on the information available in that particular format.

We also carried out tests using nutrition information presented on food packages to gain a more complete picture of the process of using nutrition information. We felt it would be likely that people would be influenced by additional information on packages, such as nutrition and health claims, or images of the food, as well as by nutrition information, and their prior knowledge of the food.

---

* Macdonald-Ross, M. (1977). How numbers are shown: a review of research on the presentation of quantitative data in texts. *Audio-visual communication review*, 25, 359-409.

Additionally in Study 1a we tested use of nutrition information when people had been given introductory summaries, explaining the main issues underlying nutrition and health, and introducing the principles of nutrition labelling.

The people tested

We tested women (social class C1) between the ages of 25 and 45 who were the principal decision-makers about shopping, and principal shoppers for their household (see details of recruitment procedure under Participants, in section 4.2 and recruitment questionnaire in Appendix 2). All participants had a general understanding of nutrition and health, which they demonstrated by making correct recommendations for healthy eating. But they were then screened further, to separate out individuals with different levels of interest in nutrition issues: a 'standard' and an 'interested' group. The screening was based on participants' own declarations about their interest in nutrition issues, and their account of how often they consulted nutrition information. While the screened groups had similar understanding of nutrition issues, it was likely that participants in the interested group were more adept than the standard group at handling the kinds of nutrition information available to consumers. This adeptness may have been a result of using nutrition information more often (because of their higher level of interest), or may have been the cause of their tendency to use nutrition information (if they found it relatively easy and, therefore, rewarding to use nutrition information, their interest in nutrition issues was likely to be sustained).

Measurements gathered

Our main measurements of the effects of the different nutrition information formats were:

- the length of time participants took to start responding (that is the length of time they took to say whether a food was a wise choice for healthy eating, or which one of two or more foods was the wiser choice)
- the length of time participants took to explain their response
- the number and quality of reasons given in their explanations.

We did not score the decisions that people made because, in many cases, there was no 'right' decision. For example, a single food might be high in sugar, but relatively low in fat, and fibre; or, in a comparison, one food might be high in fat, but also high in fibre, another low in fat, high in fibre, but also high in sugar. Individuals' decisions would depend on the priority they gave to different aspects of nutrition (for example, whether they thought cutting down on sugar was more important than cutting down on fat), and these priorities were not the focus of the study. Our interest was, given any priorities or preconceptions that people might hold, how did the different formats, and the way the formats made information available, influence factors people took into account in their decisions? So we were more interested in the reasons given in leading to decisions than in the decisions themselves.*

---

* Note, however, that we made an exception to this strategy of scoring people's reasons rather than their overall decisions. This exception arose in Study 3a, where there were frequent errors in reasoning with one format, but where it was possible that the errors in reasoning would not affect the overall decision, and so we compared the decisions made with each format (see section 8.3).

2 Summary of main findings of experimental studies

2.1 Study 1

The overall aim of the study was to examine the impact on decisions about foods of supplementing numeric information with verbal banding (high, medium, low).

Study 1a

We compared numeric information with numeric information supplemented by verbal banding, in judgements of single foods, and comparisons of two or three foods. The words high, medium or low were presented alongside numeric information. (Samples of the formats used are shown in Figure 1.) Some participants received explanatory introductions, outlining principles of nutrition and health, before carrying out the judgements tasks, others did not.

Steak & kidney pie

	per 100g	per serving 142g	
Energy	1120	1635	kJ
	267	390	kcal
Protein	9.8	14.0	g
Carbohydrate of which	25.3	37.0	g
Sugars	0.2	0.3	g
Fat of which	13.7	20.0	g
Saturated fats	5.5	7.8	g
Dietary fibre	0.3	0.4	g
Sodium	0.5	0.7	g

Steak & kidney pie

	per 100g	per serving 142g		
Energy	1120	1635	kJ	
	267	390	kcal	
Protein	9.8	14.0	g	*medium*
Carbohydrate of which	25.3	37.0	g	*medium*
Sugars	0.2	0.3	g	*low*
Fat of which	13.7	20.0	g	*high*
Saturated fats	5.5	7.8	g	*high*
Dietary fibre	0.3	0.4	g	*low*
Sodium	0.5	0.7	g	*high*

Figure 1: Numeric information, and numeric information supplemented by verbal banding, tested in Study 1a.

Main findings

For judgements of single foods banding increased the speed of response, and reduced the number of errors made, by eliminating the need for participants to interpret numeric information. For comparisons of foods it had some effects on the speed, and accuracy of reasons, (particularly for standard participants). The results suggested that even when participants had banding information, they still used numeric information, with banding as a back-up. Participants who had been classified as relatively interested in nutrition issues were better able to work between the different sources, using banding to detect large differences, and numeric information for precision.

There was no effect of introductory summaries, and we concluded that their content may have been too complex, and they may have been presented for too short a time, to have had an influence on performance.

Study 1b

We compared numeric information with banding presented without numeric information (a format that would not be allowed by the EC Nutrition Labelling Rules) in judgements of single foods, and comparisons of two foods. (Samples of the formats used are shown in Figure 2.)

Steak & kidney pie

	per 100g	per serving 142g	
Energy	1120	1635	kJ
	267	390	kcal
Protein	9.8	14.0	g
Carbohydrate of which	25.3	37.0	g
Sugars	0.2	0.3	g
Fat of which	13.7	20.0	g
Saturated fats	5.5	7.8	g
Dietary fibre	0.3	0.4	g
Sodium	0.5	0.7	g

Steak & kidney pie

	per 100g	per serving 142g	
Energy	1120	1635	kJ
	267	390	kcal
Protein	*medium*		
Carbohydrate of which	*medium*		
Sugars	*low*		
Fat of which	*high*		
Saturated fats	*high*		
Dietary fibre	*low*		
Sodium	*high*		

Figure 2: Numeric information, and verbal banding (without numeric information), tested in Study 1b.

Main findings

Responses for banding without numeric information were faster than responses for numeric information, and were faster than responses for banding as a supplement to numeric information, tested in Study 1a. This finding gave support to the conclusion that participants in Study 1a had been using both banding and numeric information when presented with banding.

For judgements of single foods, more reasons were given with banding only, because banding words were easier to use than numeric information. For comparisons, banding only reduced the number of reasons given compared to numeric information, because it obscured differences between nutrient quantities that fell within bands.

Study 1c

We compared banding presented alongside numeric information with banding presented separately from numeric information, in comparisons of two foods. (Samples of the formats used are shown in Figure 3.) The information was presented on food packaging.

Seafood tagliatelle

	per 100g	per serving 400g		
Energy	401	1682	kJ	
	95	400	kcal	
Protein	5.5	23.0	g	*high*
Carbohydrate of which	14.6	61.0	g	*medium*
Sugars	2.0	8.3	g	*medium*
Fat of which	2.0	8.5	g	*medium*
Saturated fats	1.0	4.2	g	*medium*
Dietary fibre	0.9	3.7	g	*medium*
Sodium	0.3	1.3	g	*high*

Seafood tagliatelle

	per 100g	per serving 400g	
Energy	401	1682	kJ
	95	400	kcal
Protein	5.5	23.0	g
Carbohydrate of which	14.6	61.0	g
Sugars	2.0	8.3	g
Fat of which	2.0	8.5	g
Saturated fats	1.0	4.2	g
Dietary fibre	0.9	3.7	g
Sodium	0.3	1.3	g

Protein	*high*
Carbohydrate	*medium*
Sugars	*medium*
Fat	*medium*
Saturated fats	*medium*
Dietary fibre	*medium*
Sodium	*high*

Figure 3: *Verbal banding presented alongside numeric information, and verbal banding presented separately from numeric information, tested in Study 1c.*

Main findings

Displaying banding separately from numeric information increased the time taken to respond, and reduced the number, and accuracy of reasons given in responses, compared to banding alongside numeric information. The pattern of results suggested that, ideally, banding should be presented alongside numeric information (although the constraints of packaging design sometimes rule this out).

The length of time taken to respond with banding alongside numeric information was greater here than in the equivalent condition in Study 1a, where participants saw information on paper. Furthermore participants introduced more reasons based on preconceptions about the foods they were examining here than in Study 1a. These results suggest that use of nutrition information on packages is likely to be influenced by the format of the packages, and by the presence of additional information or images on the packages.

2.2 Study 2

We compared verbal banding with a direct mapping of banding into bar charts, in comparisons of two foods. (Note that we presented verbal banding separately from numeric information here, to allow fair comparisons with bar charts, which were also presented separately from numeric information.)

Carbonara pancakes

	per 100g	per serving 114g	
Energy	780	889	kJ
	186	212	kcal
Protein	8.7	9.9	g
Carbohydrate	21.0	23.9	g
of which **Sugars**	2.7	3.1	g
Fat	8.0	9.1	g
of which **Saturated fats**	2.9	3.3	g
Dietary fibre	0.9	1.0	g
Sodium	0.5	0.6	g

Protein	*high*
Carbohydrate	*low*
Sugars	*low*
Fat	*high*
Saturated fats	*high*
Dietary fibre	*low*
Sodium	*high*

Carbonara pancakes

	per 100g	per serving 114g	
Energy	780	889	kJ
	186	212	kcal
Protein	8.7	9.9	g
Carbohydrate	21.0	23.9	g
of which **Sugars**	2.7	3.1	g
Fat	8.0	9.1	g
of which **Saturated fats**	2.9	3.3	g
Dietary fibre	0.9	1.0	g
Sodium	0.5	0.6	g

Protein	*medium*
Carbohydrate	*medium*
Sugars	*low*
Fat	*medium*
Saturated fats	*medium*
Dietary fibre	*low*
Sodium	*high*

Carbonara pancakes

	per 100g	per serving 114g	
Energy	780	889	kJ
	186	212	kcal
Protein	8.7	9.9	g
Carbohydrate	21.0	23.9	g
of which **Sugars**	2.7	3.1	g
Fat	8.0	9.1	g
of which **Saturated fats**	2.9	3.3	g
Dietary fibre	0.9	1.0	g
Sodium	0.5	0.6	g

Protein	*medium high*
Carbohydrate	*medium low*
Sugars	*low*
Fat	*medium high*
Saturated fats	*medium high*
Dietary fibre	*low*
Sodium	*high*

Carbonara pancakes

	per 100g	per serving 114g	
Energy	780	889	kJ
	186	212	kcal
Protein	8.7	9.9	g
Carbohydrate	21.0	23.9	g
of which **Sugars**	2.7	3.1	g
Fat	8.0	9.1	g
of which **Saturated fats**	2.9	3.3	g
Dietary fibre	0.9	1.0	g
Sodium	0.5	0.6	g

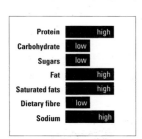

Carbonara pancakes

	per 100g	per serving 114g	
Energy	780	889	kJ
	186	212	kcal
Protein	8.7	9.9	g
Carbohydrate	21.0	23.9	g
of which **Sugars**	2.7	3.1	g
Fat	8.0	9.1	g
of which **Saturated fats**	2.9	3.3	g
Dietary fibre	0.9	1.0	g
Sodium	0.5	0.6	g

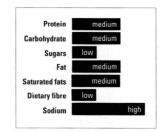

Carbonara pancakes

	per 100g	per serving 114g	
Energy	780	889	kJ
	186	212	kcal
Protein	8.7	9.9	g
Carbohydrate	21.0	23.9	g
of which **Sugars**	2.7	3.1	g
Fat	8.0	9.1	g
of which **Saturated fats**	2.9	3.3	g
Dietary fibre	0.9	1.0	g
Sodium	0.5	0.6	g

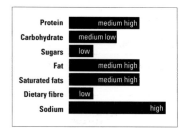

Figure 4: *Verbal banding, and bar charts, for seven nutrients, tested in Study 2.*

Different numbers of bands were compared: two (high, low), three (high, medium, low), four (high, medium-high, medium-low, low). We compared banding for all seven nutrients in the eight nutrients listed in the EC Directive (protein, carbohydrate, sugars, fat, saturated fats, dietary fibre, sodium) with banding for five nutrients (sugars, fat, saturated fats, dietary fibre, sodium). (Samples of the formats used are shown in Figures 4 & 5.)

Carbonara pancakes

	per 100g	per serving 114g	
Energy	780	889	kJ
	186	212	kcal
Protein	8.7	9.9	g
Carbohydrate of which	21.0	23.9	g
Sugars	2.7	3.1	g
Fat of which	8.0	9.1	g
Saturated fats	2.9	3.3	g
Dietary fibre	0.9	1.0	g
Sodium	0.5	0.6	g

Sugars	*low*
Fat	*high*
Saturated fats	*high*
Dietary fibre	*low*
Sodium	*high*

Carbonara pancakes

	per 100g	per serving 114g	
Energy	780	889	kJ
	186	212	kcal
Protein	8.7	9.9	g
Carbohydrate of which	21.0	23.9	g
Sugars	2.7	3.1	g
Fat of which	8.0	9.1	g
Saturated fats	2.9	3.3	g
Dietary fibre	0.9	1.0	g
Sodium	0.5	0.6	g

Sugars	*low*
Fat	*medium*
Saturated fats	*medium*
Dietary fibre	*low*
Sodium	*high*

Carbonara pancakes

	per 100g	per serving 114g	
Energy	780	889	kJ
	186	212	kcal
Protein	8.7	9.9	g
Carbohydrate of which	21.0	23.9	g
Sugars	2.7	3.1	g
Fat of which	8.0	9.1	g
Saturated fats	2.9	3.3	g
Dietary fibre	0.9	1.0	g
Sodium	0.5	0.6	g

Sugars	*low*
Fat	*medium high*
Saturated fats	*medium high*
Dietary fibre	*low*
Sodium	*high*

Carbonara pancakes

	per 100g	per serving 114g	
Energy	780	889	kJ
	186	212	kcal
Protein	8.7	9.9	g
Carbohydrate of which	21.0	23.9	g
Sugars	2.7	3.1	g
Fat of which	8.0	9.1	g
Saturated fats	2.9	3.3	g
Dietary fibre	0.9	1.0	g
Sodium	0.5	0.6	g

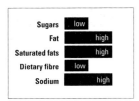

Carbonara pancakes

	per 100g	per serving 114g	
Energy	780	889	kJ
	186	212	kcal
Protein	8.7	9.9	g
Carbohydrate of which	21.0	23.9	g
Sugars	2.7	3.1	g
Fat of which	8.0	9.1	g
Saturated fats	2.9	3.3	g
Dietary fibre	0.9	1.0	g
Sodium	0.5	0.6	g

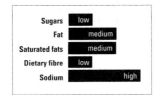

Carbonara pancakes

	per 100g	per serving 114g	
Energy	780	889	kJ
	186	212	kcal
Protein	8.7	9.9	g
Carbohydrate of which	21.0	23.9	g
Sugars	2.7	3.1	g
Fat of which	8.0	9.1	g
Saturated fats	2.9	3.3	g
Dietary fibre	0.9	1.0	g
Sodium	0.5	0.6	g

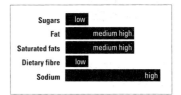

Figure 5: *Verbal banding, and bar charts, for five nutrients, tested in Study 2.*

Main findings

Bar charts reduced the time participants with a standard level of interest in nutrition issues took to start their responses compared to verbal banding, but had no effect on the number of reasons given. They had no impact on the performance of participants who had been classified as relatively interested in nutrition issues.

The different numbers of bandings did not influence responses, except for interested participants, who responded more quickly and gave fewer reasons in their responses with four-level banding than with two- or three-level banding. Further analysis showed that fewer of the interested participants' reasons were based on numeric data with four-level banding than in other conditions. The findings suggested that interested participants made an early decision that four-level banding could be used to discriminate between foods without recourse to numeric information. Two-level banding was unpopular with all participants because it showed only broad distinctions between foods.

Banding five rather than seven nutrients significantly reduced the time taken to respond, but also reduced the number of reasons given in responses. Participants preferred banding to be displayed for seven nutrients.

2.3 Study 3

Study 3 examined different graphic representations of banding:

- evaluative representations, where, for example, levels of nutrients considered beneficial to health are signalled by more, longer, or darker graphic signals, and levels not considered beneficial are signalled by fewer, shorter, or lighter graphic signals.

- direct mapping systems, where, for example, high levels are signalled by more, longer, or darker graphic signals, and low levels by fewer, shorter, or lighter graphic signals.

Study 3a

We compared verbal banding, verbal banding supplemented by a star system evaluating the levels of nutrients, and a star system that evaluated levels of nutrients (without verbal banding). (Samples of the formats used are shown in Figure 6.) The different information formats were tested in comparisons of two foods.

Main findings

Star systems increased the speed of responses, particularly for standard participants, but also reduced the numbers of reasons given in responses. Furthermore when stars were used without verbal banding, there were more errors in the reasons given (for all participants, but especially standard participants. Errors arose when participants mistakenly interpreted more or fewer stars as meaning more or less of a nutrient (that is, as a direct mapping), rather than a better or worse level of a nutrient (as an evaluative mapping). Note that these participants had not participated in any other studies, and so their errors were not prompted by prior experience of using a direct mapping system.

Study 3b

We compared verbal banding with verbal banding supplemented by bar charts, and verbal banding supplemented by a direct mapping system using shaded boxes. (Samples of the formats used are shown in Figure 7.) The different information formats were tested in comparisons of two foods.

Rice pudding

	per 100g	per serving 212g	
Energy	306	646	kJ
	72	152	kcal
Protein	3.7	7.8	g
Carbohydrate of which	12.4	26.3	g
Sugars	4.5	9.5	g
Fat of which	1.2	2.5	g
Saturated fats	0.8	1.5	g
Dietary fibre	0.2	0.4	g
Sodium	trace	trace	

Sugars	*high*
Fat	*low*
Saturated fats	*medium*
Dietary fibre	*low*
Sodium	*low*

Rice pudding

	per 100g	per serving 212g	
Energy	306	646	kJ
	72	152	kcal
Protein	3.7	7.8	g
Carbohydrate of which	12.4	26.3	g
Sugars	4.5	9.5	g
Fat of which	1.2	2.5	g
Saturated fats	0.8	1.5	g
Dietary fibre	0.2	0.4	g
Sodium	trace	trace	

The more stars shown for each nutrient, the healthier this food is for you

Sugars	high	
Fat	low	★★
Saturated fats	medium	★
Dietary fibre	low	
Sodium	low	★★

Rice pudding

	per 100g	per serving 212g	
Energy	306	646	kJ
	72	152	kcal
Protein	3.7	7.8	g
Carbohydrate of which	12.4	26.3	g
Sugars	4.5	9.5	g
Fat of which	1.2	2.5	g
Saturated fats	0.8	1.5	g
Dietary fibre	0.2	0.4	g
Sodium	trace	trace	

The more stars shown for each nutrient, the healthier this food is for you

Sugars	
Fat	★★
Saturated fats	★
Dietary fibre	
Sodium	★★

Figure 6: *Verbal banding, evaluative star system with verbal banding, and evaluative star system without verbal banding, tested in Study 3a.*

Rice pudding

	per 100g	per serving 212g	
Energy	306	646	kJ
	72	152	kcal
Protein	3.7	7.8	g
Carbohydrate of which	12.4	26.3	g
Sugars	4.5	9.5	g
Fat of which	1.2	2.5	g
Saturated fats	0.8	1.5	g
Dietary fibre	0.2	0.4	g
Sodium	trace	trace	

Sugars	*high*
Fat	*low*
Saturated fats	*medium*
Dietary fibre	*low*
Sodium	*low*

Rice pudding

	per 100g	per serving 212g	
Energy	306	646	kJ
	72	152	kcal
Protein	3.7	7.8	g
Carbohydrate of which	12.4	26.3	g
Sugars	4.5	9.5	g
Fat of which	1.2	2.5	g
Saturated fats	0.8	1.5	g
Dietary fibre	0.2	0.4	g
Sodium	trace	trace	

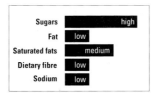

Rice pudding

	per 100g	per serving 212g	
Energy	306	646	kJ
	72	152	kcal
Protein	3.7	7.8	g
Carbohydrate of which	12.4	26.3	g
Sugars	4.5	9.5	g
Fat of which	1.2	2.5	g
Saturated fats	0.8	1.5	g
Dietary fibre	0.2	0.4	g
Sodium	trace	trace	

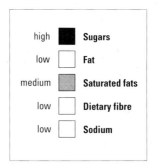

Figure 7: *Verbal banding, bar charts, and shaded boxes, tested in Study 3b.*

Main findings

Participants took longer to respond and gave fewer reasons in their responses with bar charts and shaded boxes than with verbal banding. Participants gave fewer reasons in their responses with shaded boxes than with bar charts. These results, which seem to contradict previous results by showing that graphic representations slow down responses, may have been a consequence of transfer effects from Study 3a, which this study followed, with the same participants. The transfer effects illustrate the difficulties that might be caused by having different kinds of systems representing nutrition information operating simultaneously.

2.4 Study 4

We compared different ways of representing nutrition information in relation to dietary reference values (DRVs): a numeric listing of DRVs alongside numeric information; a listing of the percentage of the DRV for each nutrient supplied by a serving of the food; a listing of percentages of DRVs, supplemented by a graphic representation. (Samples of the formats used are shown in Figure 8.) The different information formats were tested in judgements of single foods, and comparisons of two foods.

In a secondary task, participants were asked to make decisions about foods on the basis of nutrition information presented on packages. The nutrition information included both dietary reference information and banding, and we recorded the choices people made between different sources of information.

Main findings

Participants with a standard level of interest in nutrition issues responded more quickly with graphic representations of percentage DRVs than with numeric listings, or percentage listings, without the poor consequences for the number or accuracy of the reasons given that we had observed in Study 3a. Standard participants performed particularly poorly with percentage listings. Overall, participants who were relatively interested in nutrition issues took longer to respond and gave more reasons in their responses than standard participants. They were less affected by the different formats of information than standard participants.

In the de-briefing task standard participants tended to use banding information more often than numeric or percentage listings, but showed a preference for graphic representations of percentage listings compared to banding. Interested participants used banding less often, and were more likely to use numeric, percentage, or graphic representations of percentage listings.

Chicken korma

	per 100g	per serving 340g	recommended daily intake*	
Energy	921	3130	9450	kJ
	221	750	2400	kcal
Protein	12.5	45.0	75	g
Carbohydrate of which	11.0	37.0	345	g
Sugars	6.9	23.5	72	g
Fat of which	14.4	49.0	80	g
Saturated fats	7.3	24.8	27	g
Dietary fibre	4.5	15.0	30	g
Sodium	0.4	1.3	2	g
*recommended daily intake based on a 2400 kcal per day diet				

Chicken korma

	per 100g	per serving 340g		percentage of recommended daily intake* (per serving)
Energy	921	3130	kJ	
	221	750	kcal	
Protein	12.5	45.0	g	60%
Carbohydrate of which	11.0	37.0	g	11%
Sugars	6.9	23.5	g	33%
Fat of which	14.4	49.0	g	61%
Saturated fats	7.3	24.8	g	92%
Dietary fibre	4.5	15.0	g	50%
Sodium	0.4	1.3	g	65%
*recommended daily intake based on a 2400 kcal per day diet				

Chicken korma

	per 100g	per serving 340g	
Energy	921	3130	kJ
	221	750	kcal
Protein	12.5	45.0	g
Carbohydrate of which	11.0	37.0	g
Sugars	6.9	23.5	g
Fat of which	14.4	49.0	g
Saturated fats	7.3	24.8	g
Dietary fibre	4.5	15.0	g
Sodium	0.4	1.3	g

		recommended daily intake*
Protein	60%	● ● ● ● ● ● ○ ○ ○ ○
Carbohydrate	11%	● ○ ○ ○ ○ ○ ○ ○ ○ ○
Sugars	33%	● ● ● ○ ○ ○ ○ ○ ○ ○
Fat	61%	● ● ● ● ● ● ○ ○ ○ ○
Saturated fats	92%	● ● ● ● ● ● ● ● ● ○
Dietary fibre	50%	● ● ● ● ● ○ ○ ○ ○ ○
Sodium	65%	● ● ● ● ● ● ● ○ ○ ○
* the black circles show the percentage of recommended daily intake per serving (based on a 2400 kcal per day diet)		

Figure 8: Numeric dietary reference values, percentage dietary reference values, and graphic dietary reference values, tested in Study 4.

3 | Conclusions from experimental studies

3.1 Main themes arising from the research

The use of banding

Banding provides a short cut to the interpretation of numeric information, and so improves judgements of nutrient levels in single foods. It can, however, obscure distinctions between foods in comparisons if the nutrients fall within a banding level. The more fine-grained a banding system, the less likely it is to obscure distinctions. Although participants in this study recognised the value of the detail provided by a four-level banding system (high, medium-high, medium-low, low), many preferred the simplicity of a three-level system (high, medium, low).

Ideally banding should be presented alongside numeric information so that people who want to consult both banding and numeric information can do so easily.

Our finding in favour of banding contrasts with Levy and Schucker's (1991) finding. However, they tested banding only in comparisons of two foods, and faulted banding on grounds that it obscured distinctions between foods.

The use of direct and evaluative representations of banding

The participants we tested had adequate knowledge of nutrition and health to evaluate direct representations of banding where, for example, high levels of fat and fibre are represented in the same way, despite their different implications for health. A similar basic awareness of basic nutrition issues has been reported elsewhere (Charny and Lewis, 1987; Tate and Cade, 1990).* In our studies we were able to observe participants bringing this knowledge to bear on real decision-making tasks.

In contrast to the success of the direct representation of nutrient levels, we found that banding systems that evaluate levels of nutrients confuse people, so that they make errors in their judgements of foods. Similar findings are reported by Williams and Poulter (1990) in their trials of a menu labelling scheme.† Evaluative systems have the further drawback of not being applicable to all nutrients: carbohydrate levels cannot be evaluated since they include both starch and sugars.

We found further that if direct and evaluative representations of banding are used concurrently, people may not only make errors with evaluative systems, but also make errors with direct representations that they would have been unlikely to make otherwise. Consequently we

---

* Charny, M. and Lewis, P.A. (1987). Does health knowledge affect eating habits. *Health Education Journal*, 46, pp.172-6.
Tate, J. and Cade, J. (1990) Public knowledge of dietary fat and coronary heart disease. *Health Education Journal*, 49, 32-5.
† Williams, C. and Poulter, J. (1991). Formative evaluation of a workplace menu-labelling scheme. *Journal of human nutrition and dietetics*, 4, 251-262.

have suggested that only a single kind of representation should be used, preferably direct representation.

The use of graphic representations

As we discussed in section 1.2, it is impossible to give a blanket endorsement or indictment of graphic representations, since their success or failure depends on the kind of representation used. We found that bar charts representing banding, and filled circles representing percentages of DRVs were successful in speeding up the responses of standard participants, and were popular. However, using shaded boxes to represent banding led to some confusions (darker boxes were not always understood as representing higher levels of nutrients), at least in the context in which we presented them (one where participants had already used a different banding system). Although the graphic representations we tested increased the speed of standard participants' responses, they did not increase the number of reasons they gave in their responses, compared to when they used numeric or verbal information.

Our findings of some success for graphic representations appear to contradict Levy and Schucker's (1991) finding that bar charts were the least effective of a range of nutrition information formats they tested. But their bar charts were analogue representations of percentage of dietary reference values, and their participants were required to use them for comparisons across all the nutrients in pairs of foods. This is a cognitively demanding task in which the specific graphic format they used seems bound to produce inferior performance to lists of numerals presenting quantities of nutrients for comparison.

Our findings could not be used to support a strong recommendation for the use of graphic representations. But the speed with which they were approached by standard participants suggest that they might be a way of attracting people who would not normally look at nutrition information to do so.

Banding versus dietary reference values

We did not make a direct comparison between these two different ways of representing nutrition information, except in the de-briefing task of Study 4. Standard participants showed a preference for verbal banding to numeric or percentage representations of DRVs, but used graphic represenations of DRVs in preference to verbal banding. The graphic representation of DRVs we used might also be thought of as a ten-level graphic banding scheme. Since it is a direct mapping of nutrient levels, its use would be compatible with our recommendations for direct representations of banding.

It is likely that practical considerations would dictate whether or not DRVs are used: banding appears to be more versatile, since we showed it was effective whether presented verbally or graphically; dietary reference values appear to be most effective when presented graphically, and, consequently, are more limited in their application.

3.2 Putting the use of nutrition information in context

Different formats may suit different users

Throughout our studies, we found standard participants (people who do not regularly use nutrition information on food packages) were more influenced by graphic presentations (bar charts, star systems, or graphic representations of dietary reference values) than interested participants (people who claim they look at nutrition information regularly). Generally, standard participants were less likely to use numeric information than interested participants (see Study 4, especially). They had claimed, during recruitment, that they did not have any particular difficulty using information presented in grams, and there was no evidence to contradict this. However, they appeared to be less able, or willing, than interested participants to navigate different sources of information, or make quick decisions about the most appropriate information to consult.

People with lower numeracy levels than the participants we tested here might find the choice between different sources of information even more difficult to make: they might be drawn to non-numeric representations, but they might find these difficult to understand, too (Levy and Schucker, 1991). The selection of any particular representation of nutrition information will be constrained by the audience being targeted. Formats that help one set of consumers may not translate well to others with different aptitudes for handling numeric, linguistic, or graphic information.

Different formats may suit particular tasks

A further factor that must be taken into account in selecting nutrition information formats is the task that is to be performed with the information. Banding information (whether in verbal or graphic form) improved judgements of single foods by supplying support for interpretations of numeric information (Study 1a). Banding can also improve performance in comparisons of two foods, but at the expense of accuracy in the comparison (Study 1b, Study 2), since if consumers rely on banding information alone, they will miss differences between nutrient levels that fall within a band. However, since we know that people use nutrition information to make judgements about single foods, it may be better to support that task, at the expense of detail in comparisons of foods.

Individual information formats are rarely simply 'good' or 'bad'

Our research has shown that the format of nutrition information, whether numeric, verbal, or graphic can influence people's decision-making about foods. Format influenced performance most obviously in Study 3a, where an evaluative star system helped standard participants, in particular, respond more quickly, but also misled them into giving inaccurate responses. Format also influenced performance at a detailed level: moving verbal banding from alongside, to separate from numeric information, in Study 1a, reduced the speed of response, and increased the number of errors made in responding; displaying five banded nutrients, rather than seven, in Study 2, speeded up responses, but also reduced the number of reasons given in responding.

These mixed effects of different information formats demonstrate that decisions about the presentation of nutrition information are not straightforward. It is easy to rule out a system which clearly misled participants. But practical constraints may dictate that supplementary nutrition information is presented separately from numeric information (for example, when graphic representations are used). Furthermore there may be advantages in systems that elicit less detailed responses, but responses that are focussed on nutrition and health (such as a five nutrient banding system), especially if they take less time to use than a broader-based system.

4 | Report of Study 1a

4.1 Aims

The aims of Study 1a were:

- to compare people's use of numeric nutrition information (numeric format), with their use of numeric nutrition information supplemented by verbal banding (high, medium, low) (banding format) (see Figure 9), in tasks evaluating single foods and comparing more than one food.

- to compare how people with different levels of interest in nutrition issues used the two kinds of nutrition information. We tested two distinct groups of people (a 'standard' group and an 'interested' group), who had been screened before the study.

- to examine whether supplying people with a summary introducing issues currently regarded as important in nutrition and health affected use of nutrition information. We compared two summaries (one verbal, and one including numeric dietary reference values for nutrients) with a control condition where no summary was supplied.

4.2 Method

Design

The design of the study is summarised in Table 1.

Table 1: *Design of Study 1a. The number of participants in each treatment is shown in parentheses.*

Interest level of participants	Materials (set A or B) seen in numeric or banding format	Type of summary
Standard (12)	A numeric/B banding (6)	Verbal (2) Verbal + num. (2) None (2)
	B numeric/A banding (6)	Verbal (2) Verbal + num. (2) None (2)
Interested (12)	A numeric/B banding (6)	Verbal (2) Verbal + num. (2) None (2)
	B numeric/A banding (6)	Verbal (2) Verbal + num. (2) None (2)

We tested two groups of twelve people (a standard group and an interested group). Each participant saw both the numeric format and the banding format. There were two sets of materials (A and B). In each group, half of the participants saw set A in numeric format, and set B in banding format, while the other half saw set B in numeric format, and set A in banding format. And within each group the order of presentation of the different formats was balanced across participants. Within

79

each sub-group of six, two participants received a verbal summary, two received a summary including numeric dietary reference values, and two received no summary.

Materials

The labelling formats were compatible with the recommendations for the presentation of the 'Group 2' nutrients in the EC Nutrition Labelling Rules Directive (notes on the typography and graphic design of the formats are given in Appendix 1 on pp. 131-142). The formats listed energy (in kilojoules and kilocalories); protein, carbohydrate, sugars, fat, saturated fats, dietary fibre, and sodium (in grams). In the numeric format, the quantities of each nutrient were listed per 100 grams, and per serving (the serving size was given, in grams). In the banding format an appropriate banding word (high, medium, low) was shown next to the numeric information for protein, carbohydrate, sugars, fat, saturated fats, dietary fibre, and sodium. The bandings were worked out according to a scheme devised by the Coronary Prevention Group (1990), where each nutrient is banded according to its energy contribution to the total energy of the food (details of the bandings used are given in Appendix 4 on p. 146).

The labels were presented either singly or side by side on flat sheets of paper. The name of each food was given above its label (as shown in Figure 9).

An introductory summary of approximately 350 words was devised using guidelines given in *Eight guidelines for a healthy diet* (MAFF, DH, HEA, 1991).* The summary also introduced the concept of nutrition labelling. A second summary was prepared by continuing the original summary with a section giving DRVs for protein, carbohydrate, sugars, fat, saturated fats, dietary fibre, and sodium, based on a diet of 2,400 kcal per day. The summaries are shown in Appendix 3 on pp. 144-5.

Steak & kidney pie

	per 100g	per serving 142g	
Energy	1120	1635	kJ
	267	390	kcal
Protein	9.8	14.0	g
Carbohydrate	25.3	37.0	g
of which **Sugars**	0.2	0.3	g
Fat	13.7	20.0	g
of which **Saturated fats**	5.5	7.8	g
Dietary fibre	0.3	0.4	g
Sodium	0.5	0.7	g

Steak & kidney pie

	per 100g	per serving 142g		
Energy	1120	1635	kJ	
	267	390	kcal	
Protein	9.8	14.0	g	*medium*
Carbohydrate	25.3	37.0	g	*medium*
of which **Sugars**	0.2	0.3	g	*low*
Fat	13.7	20.0	g	*high*
of which **Saturated fats**	5.5	7.8	g	*high*
Dietary fibre	0.3	0.4	g	*low*
Sodium	0.5	0.7	g	*high*

Figure 9: *Numeric information, and numeric information supplemented by verbal banding, tested in Study 1a.*

* Ministry of Agriculture Fisheries and Food, Department of Health, Health Education Authority (1991). *Eight Guidelines for a Healthy Diet.* London: Food Sense.

Procedure

The participants were tested singly, in a quiet room. The procedure took about 20 minutes.

The background to the study and the procedure were described briefly to the participants.

Participants in the two groups receiving an introductory summary were asked to read the summary to themselves, at their own speed. They were told they could keep the summary by them to refer to, if necessary, during the session.

The participants made eight judgements about foods using nutrition information printed on sheets of paper. They made two judgements of single foods; then two comparative judgements of two foods, followed by a comparative judgement of three foods; the cycle of two comparisons of two foods, followed by a comparison of three foods was then repeated for a second set of materials.

Each judgement was introduced by the question 'Do you think this food would be a wise choice for a healthy diet?' (for single foods) or 'Which of these foods do you think would be a wiser choice for a healthy diet?' (for more than one food). In each case, after participants had given their initial response, they were prompted to explain their judgement. We recorded participants' initial response, and the time they took to make it; the reasons they gave in their explanation of the response, and the time taken to make their explanation.

Finally we asked questions about the tasks participants had performed:
- whether they preferred to use numeric or banding information.
- (where appropriate) whether the explanatory introduction had been useful.

Participants

We tested 24 women between the ages of 25 and 45. They were of social class C1, and were the principal decision-makers about shopping, and principal shoppers for their household. All were able to give at least three correct recommendations for healthy eating. All claimed they were 'reasonably happy' or 'quite happy' using information in grams on food packaging labels.

The participants were screened to yield two groups of 12, each with different levels of interest in nutrition issues (standard and interested), on the basis of screening questions (see Appendix 2, on p. 143). Participants whose joint rating of interest in cooking and healthy eating added up to more than 10, and who claimed to look at nutrition information frequently or always, both when shopping and at home, were classified as interested. Participants whose joint rating of interest in cooking and healthy eating added up to less than 10, and who claimed to look at nutrition information occasionally or never, were classified as standard.

Participants were recruited in Reading, Berkshire and were paid for taking part in the study.

4.3 Results

Note on introductory summaries

There were no observable effects of the different types of introductory summaries on performance. During the testing sessions only one instance was recorded where a participant looked at the summary.

Consequently we have collapsed the data across the different summary conditions. The summaries are discussed further under *Debriefing* (in this section) and in the *Discussion of results* (section 4.4).

Judgements of single foods

The mean lengths of time participants took to start and complete their responses are shown in Table 2. Participants in both the standard and interested groups started their responses more quickly when they used the banding format than when they used the numeric format (overall means 22 and 30 seconds, respectively). The difference in start times was significant for the standard group ($p < .05$). There was a trend towards longer responses with the banding format for standard participants, and shorter (but not significantly shorter) for interested participants (mean overall response lengths for numeric and banding formats were 37 and 35 seconds respectively). Although, overall, the trend was for participants in the interested group to start their responses more quickly, and to take less time to complete their responses than participants in the standard group, these differences were not significant.

Table 2: Mean times (seconds) taken to start and complete response by participants with different levels of interest, seeing different information formats in judgements of single foods in Study 1a. The range of times is shown in parentheses.

| Interest level | standard | | interested | |
Information format	numeric	banding	numeric	banding
Response start	31 (8–59)	25 (13–39)	27 (8–52)	18 (7–35)
Response length	39 (8–80)	42 (5–65)	34 (13–67)	28 (11–50)

The number of reasons given for each judgement are shown in Table 3. Participants could have given a maximum of eight reasons (since there were seven nutrients, plus energy were listed in the nutrition information for each food), but the data show that participants were selective, giving a mean of three reasons. The mean number of reasons did not differ significantly across conditions. There was no apparent difference in the number of reasons given by interested and standard participants.

Table 3: Mean number of reasons for judgements given by participants with different levels of interest, seeing different information formats in judgements of single foods in Study 1a. The range in each condition is shown in parentheses.

| Interest level | standard | | interested | |
Information format	numeric	banding	numeric	banding
Number of reasons for judgement	3 (1–5)	3 (2–5)	3 (1–4)	3 (2–5)

We analysed which nutrients participants mentioned as they explained the reasons for their judgements. No nutrient was ignored. There were some changes in the orderings of nutrients across conditions. Energy was mentioned less frequently (5 times out of a possible 24) with banding than with numeric information (9 times out 24). Conversely sodium was mentioned more frequently with banding (9 times out of 24) than with numeric information (5 times out of 24). However, other nutrients were mentioned frequently in all conditions (for example, fat had 15 out of 24 possible mentions with numeric information and 19 with banding).

The error rates in the different conditions are shown in Table 4. Errors were scored when participants' reasons for their judgements included incorrect claims about the level of a nutrient in a food. For example, they may have claimed that the level of sodium was low, when, by the banding system adopted for the study, it was high; or, conversely, they may have claimed that the level of fibre was low when, by the banding system, it was high. (Errors were not scored for participants' claims about the energy levels of the foods, since the banding system used did not define bands for energy content). There were significantly more errors when subjects saw numeric formats ($p < .05$) than when they saw banded formats (overall 35% compared to 12%). Although there was a trend for participants in the standard group to make more errors than participants in the interested group (27% compared to 19%), this was not significant.

Table 4: *Reasons where participants made errors, shown as a percentage of the total number of reasons given in each condition, in judgements of single foods. The number represented by each percentage is shown in parentheses.*

Interest level	standard		interested	
Information format	*numeric*	*banding*	*numeric*	*banding*
Errors	415 (12)	12% (4)	27% (9)	11% (4)

Participants sometimes gave additional reasons for their judgements that could not be related to individual nutrients. For example they might say they thought hot pot was a wise choice for a healthy diet because 'it's nutritious'. These reasons related to people's preconceptions about particular foods, and are covered in more detail in the discussion of results in Study 1c (section 6.4). Participants gave reasons based on preconceptions in nine (19%) of the trials; they were not given more frequently in any one condition than any other.

Comparisons of two foods

The length of time participants took to start and complete their responses is shown in Table 5. Participants in both the standard and interested groups started their responses more quickly when they used the banding format than when they used the numeric format (overall mean 24 seconds, compared to 30 seconds), although these differences were not significant. The response length was significantly shorter ($p < .01$) for interested participants (overall means 35 and 32 seconds respectively for numeric and banding). Although, overall, participants in the interested group started their responses more quickly, and took less time to complete their responses than participants in the standard group, these differences were not significant.

Table 5: *Mean times (seconds) taken to start and complete responses by participants with different levels of interest, seeing different information formats in comparisons of two foods. The range of times is shown in parentheses.*

Interest level	standard		interested	
Information format	*numeric*	*banding*	*numeric*	*banding*
Response start	32 (9–61)	25 (5–37)	26 (16–40)	23 (16–37)
Response length	32 (12–61)	34 (21–61)	37 (23–50)	29 (17–49)

The number of reasons given in each comparison of foods (maximum eight) are shown in Table 6. There was a trend for more reasons to be given using the banding format than the numeric format by both standard and interested participants, although the trend was not significant. There was also a trend towards more reasons being given by interested than by standard participants, but this was not significant.

Table 6: Mean number of reasons given in comparison of two foods by participants with different levels of interest, seeing different information formats in Study 1a. The range in each condition is shown in parentheses.

Interest level	standard		interested	
Information format	numeric	banding	numeric	banding
Number of reasons in comparison	3 (1–5)	4 (2–5)	3 (3–6)	4 (2–6)

The nutrients mentioned by participants as they explained the reasons for their judgements were analysed. There were no clear differences in the nutrients mentioned across the different formats.

The error rates in the different conditions are shown in Table 7. Errors were scored when participants' reasons included incorrect claims about the relative levels of nutrients or energy across foods. (Note there were proportionately fewer errors in this comparison task than in the judgements of single foods, since participants were making direct comparisons between parallel sets of numeric information or banding information, rather than assessing whether or not numeric information represented high, medium or low quantities of particular nutrients.) There was a trend for participants in the standard group to make more errors with the numeric than the banding format, but not the interested group who, overall, made very few errors.

Table 7: Reasons where participants made errors, shown as a percentage of the total number of reasons given in each condition, in comparisons of two foods in Study 1a. The number represented by each percentage is shown in parentheses.

Interest level	standard		interested	
Information format	numeric	banding	numeric	banding
Errors	13% (9)	5% (4)	3% (3)	3% (2)

Participants rarely gave additional reasons for their judgements that could not be related to individual nutrients (reasons relating to their preconceptions about particular foods). Reasons based on preconceptions were given in seven (7%) of the trials; they were not given more frequently in any one condition than any other. It is worth noting that on six of those trials, they involved the nutrition information for a brand of chicken korma, which, according to its nutrition information, was high in fat and saturated fat (because of the korma sauce). This was at odds with participants' expectation that chicken dishes would be low in fat.

Comparisons of three foods

The length of time participants took to start and complete their responses is shown in Table 8. Participants in both the standard and interested groups started their responses more quickly when they used the banding format than when they used the numeric format (overall means 36 and 42 seconds respectively). The response length was significantly shorter, overall, when participants used the banding format rather than the numeric format (overall means 27 and 40 seconds, respectively p < .01). There were no systematic differences in times taken by participants in the standard and interested groups to start or complete their responses. The lengths of time taken to start responses when comparing three foods were significantly longer than when comparing two foods (overall mean 32 seconds for three foods, compared to 26 seconds for two foods, p < .01). However the lengths of response in the two tasks (overall mean 34 seconds for three foods, compared to 33 seconds for two foods) did not differ significantly.

Table 8: Mean times (seconds) taken to start and complete response by participants with different levels of interest, seeing different information formats in comparisons of three foods in Study 1a. The range of times is shown in parentheses.

Interest level	standard		interested	
Information format	numeric	banding	numeric	banding
Response start	41 (13–94)	40 (18–56)	42 (23–85)	31 (19–48)
Response length	37 (13–74)	26 (9–45)	43 (15–65)	29 (16–51)

The number of reasons given in comparisons of three foods are shown in Table 9. There was no clear trend for more reasons to be given with either the numeric or banding format, although interested participants tended to give more reasons than standard participants. There was no significant difference between the mean number of reasons given here, when participants compared three foods, and when participants compared two foods.

Table 9: Mean number of reasons given in comparisons of three foods by participants with different levels of interest, seeing different information formats in Study 1a. The range in each condition is shown in parentheses.

Interest level	standard		interested	
Information format	numeric	banding	numeric	banding
Number of reasons in comparison	3 (1–5)	3 (1–5)	4 (0–6)	4 (2–5)

The nutrients mentioned by participants as they explained the reasons for their judgements were analysed. There were no clear changes in the nutrients mentioned across the different information formats, or different levels of interest.

The error rates in the different conditions are shown in Table 10. As with comparisons of two foods, errors were scored when participants' reasons in their comparisons included incorrect claims about the relative levels of nutrients, or levels of energy, across the foods. There were few errors in this comparison task, where participants were making direct comparisons between parallel sets of numeric information or numeric and banding information. There was a trend for participants in the standard group to make more errors with the numeric than the banding format. Participants in the interested group made no errors.

Table 10: Reasons where participants made errors, shown as a percentage of the total number of reasons given in each condition, in comparisons of two foods in Study 1a. The number represented by each percentage is shown in parentheses.

Interest level	standard		interested	
Information format	numeric	banding	numeric	banding
Errors	10% (4)	3% (1)	0% (0)	0% (0)

Participants rarely gave additional reasons for their judgements that could not be related to individual nutrients. Reasons based on preconceptions were given in 11 (23%) of the trials; and in all conditions.

De-briefing

Participants were asked whether, based on the task they had just carried out, nutrition information should be presented as numeric or verbal banding information. Twenty (out of twenty-four) thought that verbal banding should be used, ten saying that it should only be given as a supplement to numeric information. One participant said that using the verbal information 'felt like cheating'. Three participants said that only numeric information should be given (one saying that banding meant 'misleading approximations'). One participant thought that it would make no difference whether banding or numeric information was given.

Participants who had received introductory summaries were asked whether the summaries had been helpful in the judgement tasks they had carried out. Six (out of eight) in the standard group and seven (out of eight) in the interested group said they thought the summaries had been some help. Two (in the interested group) said that they had not previously known the carbohydrate/sugar relationship explained in the summary. One participant (in the standard group) said that the information in the summary was too difficult to remember in the decision-making task (even though participants could have looked at the summaries during the tasks).

4.4 Discussion of results in Study 1a

Effects of banding

There was a trend towards more efficient performance in judgement tasks when participants saw banding information as a supplement to numeric information, rather than numeric information on its own. With banding, participants:

- started their responses more quickly (particularly in comparisons of two foods or three foods).
- took less time to make their responses in comparison tasks.
- tended to give more reasons for their judgements of single foods and comparisons of two foods.
- made far fewer errors in their judgements of single foods.

The impact of banding was greatest in the judgement of single foods. This was not surprising, given that judging single foods without banding relied on participants' being able to interpret quantities of nutrients, and that participants were unlikely to be able to do this (see *Interpretation of numeric information*, in section 1.2). In comparisons, participants had only to pick out differences among foods, without making interpretations.

The effects of banding on judgements of single foods were also evident at a detailed level in the reasons given for decisions. The reasons given in all the decision-making tasks depended on the nutrient levels in the tested foods, and on participants' concerns about particular nutrients. In comparison tasks, the reasons given tended to be similar across conditions, because contrasts in nutrient levels would have been detected either in the numeric information or in the banding. But in judgements of single foods, sodium was mentioned comparatively infrequently when information was represented numerically, compared to when it was represented in banding. This was likely to have been because participants interpreted the relatively low quantities of sodium

86

(in grams) as low levels. With banding, the relationship between quantity and level was explicit, and so sodium was more likely to be mentioned.

We might also have expected that banding would have been helpful in comparisons of three foods, which we anticipated would be more difficult than comparisons of two foods. The longer times to start responding for three foods, compared to two foods, suggested that the task was more demanding. However, only the overall effect on response length suggested that banding made any more of a contribution to this task than to judgements of two foods.

Even when participants saw banding information, they sometimes made errors, in their judgements of single foods, and less frequently in comparisons. This result suggests that when both kinds of information were available, people sometimes used numeric information in preference to banding. The responses in the de-briefing questions suggested that participants saw banding as a supplementary, rather than prime, source of information, and sometimes did not fully trust it. In order to assess how much people might be relying on numeric information, even when they had banding information available, we decided to compare numeric information with banding only (i.e. banding without any numeric information) in Study 1b.

Except in the case of single foods, participants rarely gave reasons for their judgements that were unrelated to the nutrition information presented to them. However, the number of participants showing surprise at the fat and saturated fat content of chicken korma suggested that participants were monitoring the foods they were evaluating in the comparison tasks, although possibly only commenting when their expectations about nutrition content were violated. We felt that interpretation would be likely to play a role in more realistic tasks involving real food packages, rather than flat paper presentation. We also felt that the timings to start responses, and response lengths might differ when people handled real packages. Consequently we tested people's use of nutrition information presented on food packages in Study 1c.

Performance of standard and interested participants

We found no significant differences in the performance of standard and interested participants, although there was a general trend for interested participants to respond more quickly, and to give more reasons, more accurately, than standard participants. However, banding had a greater effect on the performance of interested participants which might suggest that interested participants may find it easier than standard participants to exploit banding information (perhaps by ignoring the distraction of numeric information, or by working effectively between the two information sources).

The effect of introductory summaries

The knowledge (accurate or inaccurate) that people bring to bear on their choices about foods is likely to have been assimilated from many different sources, over a period of years. A brief, technical summary of the kind provided in this study, would be unlikely to have any effect on performance, unless participants with summaries including dietary reference values had referred to them in their judgements of numeric information. But only one participant did so. Using the summaries in more extended training aimed at helping people understand nutrition information might have had an impact on performance, but there was no scope for such training in this study.

5 Report of Study 1b

5.1 Aims

The aims of Study 1b were:

- to compare people's use of numeric formats with their use of verbal banding (high, medium, low) without numeric information (banding only format) (see Figure 10), in evaluations of single foods and comparisons of two foods.
- to compare how people with different levels of interest in nutrition issues used the two kinds of nutrition information.

5.2 Method

Design

The design of the study is summarised in Table 11. We tested two groups of eight people (a standard group and an interested group). Each participant saw the numeric format and the banding only format. There were two sets of materials (C and D). In each group, half of the participants saw set C in numeric format, and set D in banding only format, while the other half saw set D in numeric format, and set C in banding only format. Within each group the order of presentation of the different formats was balanced across participants.

Table 11: *Design of Study 1b. The number of participants in each condition is shown in parentheses.*

Interest level of participants	Materials (set C or D) seen in numeric or banding only format
Standard (8)	C numeric/ D banding only (4)
	D numeric/ C banding only (4)
Interested (8)	C numeric/ D banding only (4)
	D numeric/ C banding only (4)

Materials

The labelling format of the numeric information was the same as in Study 1a. The banding only format was prepared by substituting banding words for all the quantities of nutrients given in the numeric formats, except the quantities of kilojoules and kilocalories, which were not altered (details of the banding scheme used are given in Appendix 4). This format would not be compatible with the provisions for the presentation of the 'Group 2' nutrients in the EC Nutrition Labelling Rules Directive (notes on the typography and graphic design of the formats are given in Appendix 1 on pp. 131-142).

The labels were presented either singly or side by side on flat sheets of paper. The name of each food was given above its label (as shown in Figure 10).

Steak & kidney pie

	per 100g	per serving 142g	
Energy	1120	1635	kJ
	267	390	kcal
Protein	9.8	14.0	g
Carbohydrate	25.3	37.0	g
of which **Sugars**	0.2	0.3	g
Fat	13.7	20.0	g
of which **Saturated fats**	5.5	7.8	g
Dietary fibre	0.3	0.4	g
Sodium	0.5	0.7	g

Steak & kidney pie

	per 100g	per serving 142g	
Energy	1120	1635	kJ
	267	390	kcal
Protein	*medium*		
Carbohydrate	*medium*		
of which **Sugars**	*low*		
Fat	*high*		
of which **Saturated fats**	*high*		
Dietary fibre	*low*		
Sodium	*high*		

Figure 10: *Numeric information, and verbal banding only (without numeric information), tested in Study 1b.*

Participants were shown the summary of nutrition issues used in Study 1a (the version without dietary reference values) before undertaking the decision-making tasks. The full summary (including dietary reference values) is shown in Appendix 3 on pp. 144-5.

Procedure

The participants were tested singly, in a quiet room. The procedure took about 15 minutes.

The background to the study and the procedure were outlined to the participants and they were asked to read the introductory summary to themselves, at their own speed. They were told they could keep the summary by them to refer to, if necessary, during the session.

The participants made six judgements about foods using nutrition information printed on sheets of paper. Their first two judgements were of single foods; they then made four comparative judgements of two foods. The administration of the task and recording of responses paralleled Study 1a, see *Procedure*, in section 4.2.

Participants

We tested 16 women, recruited using the same questionnaire as in Study 1a (see Appendix 2 on p. 145). The participants were screened to yield two groups of eight participants: standard and interested, see *Participants*, in section 4.2.

5.3 Results

Judgements of single foods

The lengths of time participants took to start and complete their response are shown in Table 12.

Participants in both the standard and interested groups tended to start their responses slightly more quickly when they used the banding only format than when they used the numeric format, although this trend

Table 12: *Mean times (seconds) taken to start and complete responses by participants with different levels of interest, seeing different information formats in judgements of single foods in Study 1b. The range of times is shown in parentheses.*

Interest level Information format	standard		interested	
	numeric	*banding only*	*numeric*	*banding only*
Response start	31 (22–41)	30 (22–41)	30 (20–41)	28 (22–33)
Response length	37 (23–62)	24 (17–33)	27 (16–39)	23 (19–32)

was not significant (overall means 29 seconds with banding only, 31 seconds with numeric information). However the response length was significantly shorter with the banding only format for standard participants ($p < .01$), and shorter (but not significantly) for interested participants (overall means 24 seconds for banding only, and 33 seconds for numeric information). There were no significant differences, overall, in the length of time taken to start or complete responses by participants in the interested group compared to participants in the standard group.

We compared the start times and response lengths here with the equivalent data from Study 1a (where participants saw numeric information and numeric information supplemented by banding). The two sets of data are shown together in Table 13. There were no significant differences in the start times or response lengths across the two studies when participants saw numeric information. However their start time was significantly longer ($p < .01$) and their response time significantly shorter ($p < .05$) when they saw banding only compared to when they saw banding used as a supplement to numeric information.

Table 13: *Comparison of mean times (seconds) taken to start and complete response by participants seeing different information formats in judgements of single foods in Study 1a and Study 1b. The range of times is shown in parentheses.*

Information format	*numeric*	*banding*
Study 1a (numeric v. numeric supplemented by banding):		
Response start	29 (8–59)	25 (7–39)
Response length	36 (13–80)	35 (5–65)
Study 1b (numeric v. banding only):		
Response start	30 (20–41)	29 (22–41)
Response length	32 (16–62)	24 (17–33)

The number of reasons given for each judgement about the relative healthiness of the foods in Study 1b are shown in Table 14. Overall, more reasons were given when people saw banding only formats than when they saw numeric formats ($p < .05$). There was a trend towards interested participants giving more reasons than standard participants, but this was not significant.

Table 14: *Mean number of reasons for judgements given by participants with different levels of interest, seeing different information formats in judgements of single foods in Study 1b. The range of times is shown in parentheses.*

Interest level Information format	standard		interested	
	numeric	*banding only*	*numeric*	*banding only*
Number of reasons for judgement	3 (2–5)	4 (3–5)	4 (3–5)	4 (3–6)

We compared the number or reasons given here with the equivalent data from Study 1a (where participants saw numeric information and numeric information supplemented by banding). The two sets of data are shown together in Table 15. There were no significant differences in the number of reasons given in the two studies when participants saw numeric information. However significantly more reasons were given when participants saw banding only compared to numeric information supplemented by banding information (p < .01).

Table 15: Comparison of mean number of reasons given for responses by participants seeing different information formats in mudgements of single foods in Study 1a and Study 1b. The range of number of reasons is shown in parentheses. (Note that the number of reasons are shown to one decimal place here, in order to clarify differences between conditions.)

Information format	numeric	banding
Study 1a (numeric v. numeric supplemented by banding): Number of reasons for judgement	3.2 (1–5)	3.3 (2–5)
Study 1b (numeric v. banding only): Number of reasons for judgement	3.6 (2–5)	4.3 (3–6)

The nutrients mentioned by participants as they explained the reasons for their judgements were analysed. There were some changes in the nutrients mentioned in different conditions. Saturated fat was mentioned in 12 out of 16 trials in banding only format compared to 5 out of 16 trials with numeric format. Sodium was mentioned in 8 out of 16 trials in the banding only format, compared to 1 out of 16 trials in the numeric format.

The error rates in the different conditions are shown in Table 16. (As in Study 1a, errors were not scored for participants' claims about the energy levels of the foods.) Participants made errors only when they used numeric formats, but not when they used banding only formats. (See, for comparison, the higher number of errors made in judgements of single foods in Study 1a, including a small number of errors when banding information was used with numeric information, Table 4, in section 4.3.)

Table 16: Reasons where participants made errors, shown as a percentage of the total number of reasons given in each condition, in judgements of single foods. The number represented by each percentage is shown in parentheses.

Interest level Information format	standard		interested	
	numeric	banding only	numeric	banding only
Errors	26% (6)	0% (0)	14% (4)	0% (0)

Participants gave additional reasons for their judgements that could not be related to individual nutrients in eight (25%) of the trials; seven of these reasons were given when participants used numeric formats, and one when they used a banding only format.

Comparisons of two foods

The length of time participants took to start and complete their responses is shown in Table 17. There were non-significant trends for participants to start their responses more quickly for banding only than for numeric information (overall means 26 and 28 seconds respectively), and to give shorter responses when they used banding only rather than numeric information (overall means 25 and 28 seconds respectively). There were no overall differences in the start times or response lengths for standard and interested participants.

Table 17: Mean times (seconds) taken to start and complete response by participants with different levels of interest, seeing different information formats in comparisons of two foods. The range of times is shown in parentheses.

Interest level	standard		interested	
Information format	numeric	banding only	numeric	banding only
Response start	28 (24–33)	27 (23–30)	28 (21–32)	25 (20–29)
Response length	28 (20–32)	24 (20–27)	28 (22–35)	26 (19–44)

We compared the start times and response lengths here with the equivalent data from Study 1a (where participants saw numeric information and numeric information supplemented by banding). The two sets of data are shown together in Table 18. There were no significant differences in the start times or response lengths across the two studies when participants saw numeric information. Nor was there any difference in start time when participants saw banding information as a supplement to numeric information (Study 1a) compared to seeing it on its own (Study 1b). However, as with judgements of single foods, response lengths were significantly shorter ($p < .05$) when participants saw banding only compared to numeric information supplemented by banding.

Table 18: Comparison of mean times (seconds) taken to start and complete response by participants seeing different information formats in comparisons of two foods in Study 1a and Study 1b. The range of times is shown in parentheses.

Information format	numeric	banding only
Study 1a (numeric v. numeric supplemented by banding):		
Response start	29 (9–60)	24 (5–37)
Response length	34 (12–61)	32 (17–61)
Study 1b (numeric v. banding only):		
Response to start	28 (21–33)	26 (20–30)
Response length	29 (20–35)	25 (19–44)

The number of reasons given in each comparison of foods are shown in Table 19. Fewer reasons were given using the banding only format than the numeric format ($p < .05$). There were no significant differences between the number of reasons given by standard and interested participants.

Table 19: Mean number of reasons given in comparison of two foods by participants with different levels of interest, seeing different information formats. The range in each condition is shown in parentheses.

Interest level	standard		interested	
Information format	numeric	banding only	numeric	banding only
Number of reasons in comparison	4 (2–5)	3 (2–4)	4 (3–5)	3 (2–4)

The nutrients mentioned by participants as they explained the reasons for their judgements were analysed. There were no notable changes in the nutrients mentioned across conditions.

There were no errors in the reasons given by participants in any of the conditions.

Participants rarely gave additional reasons for their judgements that could not be related to individual nutrients (reasons relating to their preconceptions about particular foods, discussed above in the results for single foods). Reasons based on preconceptions were given in 3% of the trials, all when participants were using the numeric format.

5.4 Discussion of results in Study 1b

Effects of banding only

There was a trend towards more efficient performance in judgement tasks when participants saw banding only, rather than numeric information. With banding only, participants:

- tended to start their responses more quickly in judgements of single foods and comparisons of two foods.
- tended to take less time to complete their responses.
- tended to give more reasons with banding only than with numeric information when they were making judgements of single foods.
- tended to give different priorities to nutrients in their reasoning in judgements of single foods, compared to when they saw numeric information.

However, although participants (at least standard participants) gave significantly more reasons in their judgements of single foods with banding only, they gave significantly fewer reasons when they were comparing foods. This was because, with banding only, contrasts between quantities of nutrients within a band were obscured, and there was no numeric information to refer to.

There were no significant differences between this study and Study 1a in the timing data or number of reasons given when participants saw numeric information. However there were differences in the performance of participants with banding only, presented here, and banding as a supplement to numeric information, presented in Study 1a. Comparing equivalent tasks performed with banding only rather than banding as a supplement to numeric information, participants:

- took significantly longer to start their responses, but took significantly less time to complete their responses in judgements of single foods; and took significantly less time to start and complete their responses in comparisons of two foods.
- gave significantly more reasons in their judgements of single foods, although not in comparisons of two foods (where banding only obscured differences in quantities between foods).

This contrast in performance between Study 1a and Study 1b suggests that in Study 1a, participants were still influenced by numeric information, even when they were supplied with banding. This meant that their responses took longer, but also, in the case of comparisons between two foods, that they were fuller.

Performance of standard and interested participants

We found no significant differences in the performance of standard and interested participants. However, there were more significant differences between performance with numeric information and with banding only for standard participants than for interested participants. This confirmed the suggestion in Study 1a that standard participants may find it more difficult to choose between different sources of information (such as numeric and verbal) to find the most appropriate for a decision-making task.

6 Report of Study 1c

6.1 Aims

The aims of study 1 were:

- to examine people's use of nutrition information formats similar to those tested in Study 1a and Study 1b, but here presented on food packaging. Two different formats were examined: verbal banding presented alongside numeric information, as in Study 1a (banding alongside), and verbal banding presented in a separate, boxed display beneath the numeric information (banding separate) (see Figure 11).

- to compare how people with different levels of interest in nutrition issues used the two kinds of nutrition information.

Seafood tagliatelle

	per 100g	per serving 142g	
Energy	401	1682 kJ	
	95	400 kcal	
Protein	5.5	23.0 g	*high*
Carbohydrate	14.6	61.0 g	*medium*
of which **Sugars**	2.0	8.3 g	*medium*
Fat	2.0	8.5 g	*medium*
of which **Saturated fats**	1.0	4.2 g	*medium*
Dietary fibre	0.9	3.7 g	*medium*
Sodium	0.3	1.3 g	*high*

Seafood tagliatelle

	per 100g	per serving 142g
Energy	401	1682 kJ
	95	400 kcal
Protein	5.5	23.0 g
Carbohydrate	14.6	61.0 g
of which **Sugars**	2.0	8.3 g
Fat	2.0	8.5 g
of which **Saturated fats**	1.0	4.2 g
Dietary fibre	0.9	3.7 g
Sodium	0.3	1.3 g

Protein	*high*
Carbohydrate	*medium*
Sugars	*medium*
Fat	*medium*
Saturated fats	*medium*
Dietary fibre	*medium*
Sodium	*high*

Figure 11: Verbal banding presented alongside numeric information, and verbal banding presented separately from numeric information, tested in Study 1c.

94

6.2 Method

Design

The design of the study is summarised in Table 20. We tested two groups of twelve people (a standard group and an interested group). Each participant saw both the banding alongside and banding separate format. There were two sets of materials: E and F. In each group, half of the participants saw set E in banding alongside format, and set F in banding separate format, while the other half saw set E in banding separate format, and set F in banding alongside format. Within each group the order of presentation of the different formats was balanced across participants.

Table 20: Design of study 1c. The number of participants in each condition is shown in parentheses.

Interest level of participants	Materials (set E or F) seen in banding alongside or banding separate format
Standard (12)	E numeric/ F banding (6)
	F numeric/ E banding (6)
Interested (12)	E numeric/ F banding (6)
	F numeric/ E banding (6)

Materials

The labelling formats were compatible with the provisions for the presentation of the 'Group 2' nutrients in the EC Nutrition Labelling Rules Directive (notes on the typography and graphic design of the formats are given in Appendix 1, pp. 131-142). The formats to be tested were substituted for any existing nutrition labelling on the food packages used. As far as possible the colour schemes used on the packages were reproduced on the substitute labels.

Procedure

The study followed directly from Study 1a. It took about 10 minutes.

The participants carried out six comparative judgements of pairs of foods using nutrition information presented on the food packaging. In each case they were asked to say which food was the wiser choice for a healthy diet, and their responses were recorded in the same way as described in Study 1a and 1b for nutrition information presented on paper.

After they had completed the comparison tasks participants were asked:
- what they felt were the differences between looking at nutrition information on paper and on packages.
- whether they had any comments on the different formats for banding information used on the packages.

Participants

The participants were the same as in Study 1a.

6.3 Results

Judgement task

The lengths of time participants took to start and complete the comparisons are shown in Table 21.

Table 21: Mean times taken to start and complete response by participants with different levels of interest, seeing banding either alongside or separate from numeric information in comparisons of food packages in Study 1c. The range of times is shown in parentheses.

Interest level	standard		interested	
Position of banding	alongside	separate	alongside	separate
Response start	34 (17–54)	33 (21–61)	30 (22–43)	36 (20–51)
Response length	37 (23–61)	40 (19–54)	37 (22–46)	40 (28–51)

Participants in the interested group took significantly longer to start their responses when banding was presented separately, rather than alongside numeric information ($p < .01$), but there was no difference in start times for standard participants. However, overall, participants took longer to complete their responses when banding information was presented separately from numeric information, rather than alongside it (overall means 40 and 37 seconds respectively, $p < .05$). There were no significant differences, overall, between the length of time taken by participants with different levels of interest to start or complete their response.

We compared the start times and response lengths here, where participants saw banding alongside numeric information, with the equivalent condition in Study 1a (where participants saw numeric information supplemented by banding on paper, rather than on packages). The two sets of data are shown together in Table 22. The start times were significantly longer when participants saw information on packages than when they saw information on paper ($p < .01$). There was a trend toward longer response lengths with information on packages, but this was not significant.

Table 22: Comparison of mean times taken to start and complete response by participants seeing banding alongside numeric information presented on paper (Study 1a) and on food packages (Study 1c). The range of times is shown in parentheses.

Banding positioned alongside numeric information

Study 1a (information presented on paper):
Response start   24 (5–37)
Response length   32 (17–61)

Study 1c (information presented on food packages):
Response start   32 (17–54)
Response length   37 (22–51)

The numbers of reasons given in each comparison of foods are shown in Table 23.

Table 23: Mean number of reasons given in comparisons of two foods by participants with different levels of interest, seeing banding either alongside or separate from numeric information in comparisons of food packages in Study 1c. The range in each condition is shown in parentheses. (Note that the number of reasons are shown to one decimal place here in order to clarify differences between conditions.)

Interest level	standard		interested	
Position of banding	alongside	separate	alongside	separate
Number of reasons in comparison	3.2 (2–5)	3.1 (1–4)	3.3 (1–5)	2.8 (2–4)

There was a trend for fewer reasons to be given when banding was separate from numeric information rather than alongside numeric information, but this was significant only for participants in the interested group (p < .05). There were no significant differences, overall, between the number of reasons given by standard and interested participants.

We compared the number of reasons given in responses here (overall mean 3) with the number of reasons given in the equivalent condition in Study 1a, where participants saw numeric information supplemented by banding, on paper, rather than on packages, (overall mean 4). There was no significant difference between the number of reasons given in the two tasks.

There were no clear differences in the nutrients mentioned in reasons given for decisions across the different conditions.

There was a non-significant trend for participants to make more errors when they saw banding separate from numeric information than when they saw it alongside numeric information (see Table 24).

Table 24: Reasons where participants made errors (shown as a percentage of the total number of reasons given in each condition) in judgements of nutrition information for pairs of foods, presented on packages. The number represented by each percentage is shown in parentheses.

| Interest level | standard | | interested | |
Position of banding	alongside	separate	alongside	separate
Errors	10% (9)	14% (14)	1% (1)	5% (5)

In all conditions participants gave reasons for their judgements that could not be related to individual nutrients. In addition to general preconceptions (such as, 'yoghurt is healthy') participants often mentioned information other than nutrition information on a package, such as a Weightwatchers flash (for rice pudding) or a panel giving detailed calorie information (Special K), or an image suggesting healthiness (Fruit 'n Fibre). These kinds of reasons were given in 38% of the trials, more frequently than in all the studies where nutrition information was presented on paper.

De-briefing

Participants were asked whether they thought there was a difference between making judgements about foods when they saw information on paper (in Study 1a) and when they saw information presented on food packages. Six participants (three standard and three interested) said that they were influenced by information other than nutrition information when they were examining packets, but not when they saw information on paper. Two participants (in the standard group) said that they found nutrition information more difficult to read on packages (in some cases it was presented at a smaller size than on paper, so that it would fit on the package). The remaining 16 participants said they did not think there was any difference between the two tasks.

Participants were asked whether they preferred banding information presented alongside or separate from nutrition information. Eighteen (7 standard, 11 interested) preferred banding alongside numeric information. Three (in the standard group) preferred banding presented separately from nutrition information. The remaining three had no preference.

6.4 Discussion of results in Study 1c

When banding was presented separately from, rather than alongside, numeric information, participants:

- tended to take longer to start their responses.
- took significantly longer to complete their responses.
- gave fewer responses.
- tended to make more errors.

The lower level of responses with banding presented separately suggests that participants may have been treating this format like the banding only format tested in Study 1b. Coupled with the higher number of errors when banding was presented separately, and participants' preferences, expressed in de-briefing, the study suggested that presenting banding alongside numeric information provides better support for decision-making than presenting it separately. However this is not always a practical option for the presentation of graphic representations.

The finding that participants took longer when using information presented on packages rather than on paper is likely to be a consequence of their needing longer to orient themselves to nutrition information on packages, given the distraction of handling the packages, and of other information displayed on them. The relatively high number of preconceived reasons given in responses suggested that, even when examining nutrition information, people are likely to be influenced by the context in which that information is presented: whether that context is their prior knowledge of the food, or prompts from the food packaging.

7 | Report of Study 2

7.1 Aims

The aims of Study 2 were:

- to compare people's use of banding presented verbally, with banding presented in a combined verbal/bar chart format (see Figures 12 & 13, pp. 100-1).
- to compare people's use of two-level (high/low), three-level (high/medium/low) and four-level (high, medium-high, medium-low, low) banding systems (see Figures 12 & 13).
- to compare people's use of banding given for seven main nutrients with banding used for a sub-set of nutrients, considered particularly relevant to health (five nutrients) (see Figures 12 & 13).

In keeping with the EC Directive, banding was always presented as a supplement to numeric nutrition information. The different formats were all examined in tasks involving the comparison of two foods.

There were two groups of participants in the study, standard and interested. Interest level was included in the analysis of the data.

7.2 Method

Design

The design of the study is summarised in Table 25.

Table 25: *Design of Study 2. The number of participants in each condition is shown in parentheses. In each condition, half the participants were standard and half were interested.*

Banding level	Materials (set G or H) seen in verbal or bar chart format	Subset 1 or 2 banded for 7 or 5 nutrients
2-band (8)	G verbal/ H bar chart (4)	Subset 1: 7; subset 2: 5 (2) Subset 2: 7; subset 1: 5 (2)
	H verbal/ G bar chart (4)	Subset 1: 7; subset 2: 5 (2) Subset 2: 7; subset 1: 5 (2)
3-band (8)	G verbal/ H bar chart (4)	Subset 1: 7; subset 2: 5 (2) Subset 2: 7; subset 1: 5 (2)
	H verbal/ G bar chart (4)	Subset 1: 7; subset 2: 5 (2) Subset 2: 7; subset 1: 5 (2)
4-band (8)	G verbal/ H bar chart (4)	Subset 1: 7; subset 2: 5 (2) Subset 2: 7; subset 1: 5 (2)
	H verbal/ G bar chart (4)	Subset 1: 7; subset 2: 5 (2) Subset 2: 7; subset 1: 5 (2)

We tested three groups of eight people: one group saw two-level banding, one saw three-level banding, and one four-level banding. All participants saw both verbal banding (verbal format) and combined

Carbonara pancakes

	per 100g	per serving 142g	
Energy	780	889	kJ
	186	212	kcal
Protein	8.7	9.9	g
Carbohydrate	21.0	23.9	g
of which **Sugars**	2.7	3.1	g
Fat	8.0	9.1	g
of which **Saturated fats**	2.9	3.3	g
Dietary fibre	0.9	1.0	g
Sodium	0.5	0.6	g

Protein	*high*
Carbohydrate Sugars	*low* *low*
Fat Saturated fats	*high* *high*
Dietary fibre	*low*
Sodium	*high*

Carbonara pancakes

	per 100g	per serving 142g	
Energy	780	889	kJ
	186	212	kcal
Protein	8.7	9.9	g
Carbohydrate	21.0	23.9	g
of which **Sugars**	2.7	3.1	g
Fat	8.0	9.1	g
of which **Saturated fats**	2.9	3.3	g
Dietary fibre	0.9	1.0	g
Sodium	0.5	0.6	g

Protein	*medium*
Carbohydrate Sugars	*medium* *low*
Fat Saturated fats	*medium* *medium*
Dietary fibre	*low*
Sodium	*high*

Carbonara pancakes

	per 100g	per serving 142g	
Energy	780	889	kJ
	186	212	kcal
Protein	8.7	9.9	g
Carbohydrate	21.0	23.9	g
of which **Sugars**	2.7	3.1	g
Fat	8.0	9.1	g
of which **Saturated fats**	2.9	3.3	g
Dietary fibre	0.9	1.0	g
Sodium	0.5	0.6	g

Protein	*medium high*
Carbohydrate Sugars	*medium low* *low*
Fat Saturated fats	*medium high* *medium high*
Dietary fibre	*low*
Sodium	*high*

Carbonara pancakes

	per 100g	per serving 142g	
Energy	780	889	kJ
	186	212	kcal
Protein	8.7	9.9	g
Carbohydrate	21.0	23.9	g
of which **Sugars**	2.7	3.1	g
Fat	8.0	9.1	g
of which **Saturated fats**	2.9	3.3	g
Dietary fibre	0.9	1.0	g
Sodium	0.5	0.6	g

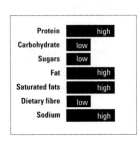

Carbonara pancakes

	per 100g	per serving 142g	
Energy	780	889	kJ
	186	212	kcal
Protein	8.7	9.9	g
Carbohydrate	21.0	23.9	g
of which **Sugars**	2.7	3.1	g
Fat	8.0	9.1	g
of which **Saturated fats**	2.9	3.3	g
Dietary fibre	0.9	1.0	g
Sodium	0.5	0.6	g

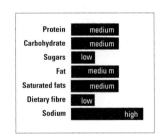

Carbonara pancakes

	per 100g	per serving 142g	
Energy	780	889	kJ
	186	212	kcal
Protein	8.7	9.9	g
Carbohydrate	21.0	23.9	g
of which **Sugars**	2.7	3.1	g
Fat	8.0	9.1	g
of which **Saturated fats**	2.9	3.3	g
Dietary fibre	0.9	1.0	g
Sodium	0.5	0.6	g

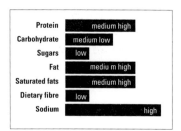

Figure 12: *Verbal banding, and bar charts, for seven nutrients, tested in Study 2.*

Carbonara pancakes

	per 100g	per serving 114g	
Energy	780	889	kJ
	186	212	kcal
Protein	8.7	9.9	g
Carbohydrate of which	21.0	23.9	g
Sugars	2.7	3.1	g
Fat of which	8.0	9.1	g
Saturated fats	2.9	3.3	g
Dietary fibre	0.9	1.0	g
Sodium	0.5	0.6	g

Sugars	*low*
Fat	*high*
Saturated fats	*high*
Dietary fibre	*low*
Sodium	*high*

Carbonara pancakes

	per 100g	per serving 114g	
Energy	780	889	kJ
	186	212	kcal
Protein	8.7	9.9	g
Carbohydrate of which	21.0	23.9	g
Sugars	2.7	3.1	g
Fat of which	8.0	9.1	g
Saturated fats	2.9	3.3	g
Dietary fibre	0.9	1.0	g
Sodium	0.5	0.6	g

Sugars	*low*
Fat	*medium*
Saturated fats	*medium*
Dietary fibre	*low*
Sodium	*high*

Carbonara pancakes

	per 100g	per serving 114g	
Energy	780	889	kJ
	186	212	kcal
Protein	8.7	9.9	g
Carbohydrate of which	21.0	23.9	g
Sugars	2.7	3.1	g
Fat of which	8.0	9.1	g
Saturated fats	2.9	3.3	g
Dietary fibre	0.9	1.0	g
Sodium	0.5	0.6	g

Sugars	*low*
Fat	*medium high*
Saturated fats	*medium high*
Dietary fibre	*low*
Sodium	*high*

Carbonara pancakes

	per 100g	per serving 114g	
Energy	780	889	kJ
	186	212	kcal
Protein	8.7	9.9	g
Carbohydrate of which	21.0	23.9	g
Sugars	2.7	3.1	g
Fat of which	8.0	9.1	g
Saturated fats	2.9	3.3	g
Dietary fibre	0.9	1.0	g
Sodium	0.5	0.6	g

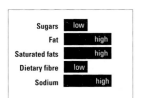

Carbonara pancakes

	per 100g	per serving 114g	
Energy	780	889	kJ
	186	212	kcal
Protein	8.7	9.9	g
Carbohydrate of which	21.0	23.9	g
Sugars	2.7	3.1	g
Fat of which	8.0	9.1	g
Saturated fats	2.9	3.3	g
Dietary fibre	0.9	1.0	g
Sodium	0.5	0.6	g

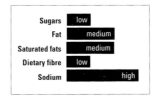

Carbonara pancakes

	per 100g	per serving 114g	
Energy	780	889	kJ
	186	212	kcal
Protein	8.7	9.9	g
Carbohydrate of which	21.0	23.9	g
Sugars	2.7	3.1	g
Fat of which	8.0	9.1	g
Saturated fats	2.9	3.3	g
Dietary fibre	0.9	1.0	g
Sodium	0.5	0.6	g

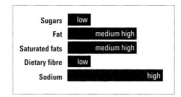

Figure 13: *Verbal banding, and bar charts, for five nutrients, tested in Study 2.*

101

verbal/bar chart (bar chart format). There were two sets of materials (G and H). In each group, half of the participants saw set G in verbal format, and set H in bar chart format, while the other half saw set H in verbal format, and set G in bar chart format. In each of the sub-groups seeing different sets of materials in different formats, half the participants saw subsets G1 and H1 with banding for seven nutrients, and G2 and H2 with banding for five nutrients; and the other half saw subsets G2 and H2 with banding for seven nutrients, and G1 and H1 with banding for five nutrients. The order of presentation of the different formats, and of the different numbers of nutrients, was balanced across participants.

Within each group seeing different formats, half the participants were classified as standard, and half as interested, according to the screening questions incorporated into the recruitment questionnaire. One standard, and one interested, participant was allotted to each of the pairs of participants seeing each sub-set of materials in each condition.

Materials

The labelling formats (shown in Figures 12 & 13) were based on the requirements for the presentation of the 'Group 2' nutrients in the EC Nutrition Labelling Rules Directive (notes on the typography and graphic design of the formats are given in Appendix 1, pp. 131-142). The formats all consisted of a panel of numeric information (the same as the numeric information presented in Studies 1a–c). Beneath the numeric panel was an additional panel giving banding in either verbal or bar chart format (details of the banding scheme used are given in Appendix 4, p. 146). The verbal format was the same as the format used for separate banding in Study 1c. The bar chart format listed the nutrients, each with a black, horizontal bar to its right, the length of the bar representing the level of nutrient in the food. The level of the nutrient was given verbally, enclosed within the bar, in white type, reversed out of the black background provided by the bar.

The findings of Experiment 1c had shown that, in some cases, people's performance in comparison tasks tended to be less effective when banding information was presented in a separate display, rather than alongside numeric information, and that people's preference was for banding alongside numeric information. However, we decided to present banding separately here because sample labels with a bar chart presented alongside numeric information seemed difficult to scan, and were judged to be so by participants in the discussion group/ questionnaire study.

The labels were presented side by side on flat sheets of paper. The name of each food was given above its label (as shown in Figures 12 & 13).

An introductory summary, as in Study 1a (the version without dietary reference values, see Appendix 3, pp. 144-5), was used.

Procedure

Study 2 followed directly from Study 1c. Participants were tested singly, in a quiet room. The procedure took about 10 minutes.

All participants were shown the introductory summary. They were asked to read it in their own time, and were told they could refer to it during the session if they wished. Participants who had seen a summary (in Study 1a) were asked to read it again to refresh their memories.

The participants made eight comparisons of two foods, using nutrition information printed on sheets of paper. In each case participants were asked to say which food was the wiser choice for a healthy diet, and their responses were recorded in the same way as described in Studies 1a–c.

Finally we asked questions about the tasks participants had performed:
- whether they preferred the verbal or bar chart format.
- (where appropriate) how using two-level or four-level banding in this study compared to the three-level banding they had used previously; whether they preferred the present version, or the one they used previously.
- if they noticed that in some cases they had been presented with seven nutrients and in some cases five nutrients; and whether they preferred either seven or five nutrients to be listed.

Participants

The participants were the same as in Studies 1a and 1c. This meant that by the time they took part in this study, they were relatively experienced in examining nutrition information.

7.3 Results

Comparison task

The mean times for standard and interested participants to start responding with each banding format and with each level of banding are shown in Table 26 (the means are combined for conditions where participants saw banding for seven nutrients and five nutrients). Standard participants took significantly longer to start their responses when presented with verbal formats compared to bar chart formats (overall means 32 seconds and 22 seconds respectively, $p < .05$). In contrast, there was a trend for interested participants to take slightly less time to start their responses when they saw verbal formats than when they saw bar chart formats (overall means 25 and 29 seconds respectively), but this difference was not significant.

There was no significant difference in start times when participants saw two- or three-level banding. However, start times were significantly shorter when participants saw four-level banding ($p < .01$). Further analysis showed that this difference was significant for interested participants (overall means for three-level and four-level banding, 36 and 23 seconds respectively, $p < .01$), but not for standard participants (overall means for three-level and four-level banding, 25 and 22 respectively).

Table 26: Mean times (seconds) taken to start responses by participants with different levels of interest, seeing verbal and bar chart formats, with banding at two-, three-, and four-levels, in Study 2. The range of times in each condition is shown in parentheses.

Levels of banding		2	3	4
Verbal format:	standard	42 (11–68)	28 (15–40)	25 (10–40)
	interested	20 (8–32)	34 (16–65)	21 (11–38)
	overall	31 (8–68)	31 (16–65)	23 (10–40)
Bar chart format:	standard	23 (10–41)	23 (6–40)	20 (5–35)
	interested	24 (11–37)	38 (19–88)	24 (7–48)
	overall	24 (10–41)	31 (6–88)	22 (5–48)

The overall start times for standard and interested participants did not vary systematically.

The start times for labelling displaying seven nutrients and labelling displaying five nutrients are shown in Table 27.

Table 27: Mean times (seconds) taken to start responses (across participants with different levels of interest, seeing verbal and bar chart formats, with banding at two-, three-, and four-levels) in comparisons of two foods in Study 2. The range of times is shown in parentheses.

Number of nutrients displayed	7	5
Times	31 (9–88)	23 (5–65)

Start times were significantly shorter when five nutrients were displayed than when seven nutrients were displayed (p<.001). This difference was significant across all conditions.

The mean length of time to complete responses with each banding format and with each level of banding is shown in Table 28 (the means are combined for conditions where participants saw banding for seven nutrients and five nutrients). There was no significant difference in response length with verbal or bar chart formats.

Overall, there was a trend towards shorter response lengths from two- to three-, to four-level banding (overall means, 36, 30 and 28 seconds respectively). The trend was significant only between three- and four-level banding for interested participants (means for three- and four-levels, 35 and 23 respectively, p<.01).

Table 28: Mean response lengths (seconds) for participants with different levels of interest, seeing verbal and bar chart formats, with banding at two-, three-, and four-levels, in Study 2. The range of times in each condition is shown in parentheses.

Levels of banding		2	3	4
Verbal format:	standard	34 (17–58)	23 (8–33)	33 (22–65)
	interested	42 (19–85)	34 (19–55)	24 (13–39)
	overall	38 (17–85)	29 (8–55)	29 (13–65)
Bar chart format:	standard	33 (19–60)	27 (9–44)	29 (15–56)
	interested	34 (16–72)	35 (23–65)	22 (11–31)
	overall	34 (16–72)	31 (9–65)	26 (1–56)

The start times and response lengths for standard and interested participants did not vary systematically.

The mean times taken to complete responses for labelling displaying seven nutrients and labelling displaying five nutrients are shown in Table 29. Response lengths were significantly shorter when five nutrients were displayed than when seven nutrients were displayed (p<.001). This difference was significant across all conditions.

Table 29: Mean times (seconds) taken to complete responses (across participants with different levels of interest, seeing verbal and bar chart formats, with banding at two-, three-, and four-levels) in Study 2. The range of times is shown in parentheses.

Number of nutrients displayed	7	5
Times	35 (11–85)	26 (8–72)

The mean number of reasons given in responses with each banding format, and with each level of banding are shown in Table 30 (the means are combined for conditions where participants saw banding for seven nutrients and five nutrients). There was no significant difference, overall, in the number of reasons given with verbal or bar chart formats. Nor did the number of reasons, overall, differ across different levels of banding. However further analysis showed that there was a significant decrease in the number of reasons given by interested participants with four-level banding, compared to three-level banding (means for four-level and three-level, 3 and 2 responses respectively, $p < .01$). Overall there were no significant differences between standard and interested participants, except for four-level banding, where interested participants gave significantly fewer reasons than standard participants (means 3 for standard and 2 for interested, $p < .05$).

Table 30: Mean number of reasons given in responses by participants with different levels of interest, seeing verbal and bar chart formats, with banding at two-, three-, and four-levels, in Study 2. The range in each condition is shown in parentheses.

Levels of banding		2	3	4
Verbal format:	standard	3 (0–5)	3 (2–5)	4 (2–5)
	interested	3 (2–5)	3 (1–4)	2 (2–4)
	overall	3 (0–5)	2 (1–5)	3 (2–5)
Bar chart format:	standard	3 (1–5)	3 (1–4)	3 (2–5)
	interested	3 (1–5)	3 (2–5)	2 (2–4)
	overall	3 (1–5)	3 (1–5)	3 (2–5)

By analysing the reasons given we could isolate those that could not have been based on the information in the banding panel, and where participants must have used the numeric information given in the label: for example, where participants said there was a difference between levels of nutrient in different foods that fell within the same band in the banding panels; or where they said there was a trace of a nutrient (which would have been shown as 'low' on the banding panels). Table 31 shows the percentage of the responses given in each condition that could only have been given on the basis of numeric information. Overall, significantly fewer reasons were derived from numeric data with four-level banding than with three-level banding ($p < .01$), or two-level banding ($p < .05$) (percentages for four-, three-, and two-level banding, 34, 63, and 51, respectively). This difference was most marked for interested participants ($p < .05$ in both comparisons of four with three levels, and of four with two levels). So interested participants were not only giving fewer responses than standard participants for four-level banding, but a smaller proportion of those responses was derived from numeric data. (Note that the trend for more reasons to be

Table 31: Percentage of reasons based on numeric information given by participants with different levels of interest, seeing verbal and bar chart formats, with banding at two-, three-, and four-levels, in Study 2.

Levels of banding		2	3	4
Verbal format:	standard	60	84	59
	interested	55	66	26
	overall	57	75	43
Bar chart format:	standard	40	39	26
	interested	50	63	23
	overall	45	51	25

derived from numeric information with three-level banding than with two-level banding was not significant.)

We isolated responses that were based on banding information only (that is, complete responses where none of the reasons given were based on numeric information). Table 32 shows that there were many more trials where participants used only banding information with four-level banding than with two-, or three-level banding (totals for two-, three-, and four-level: 8, 0, and 32 trials respectively). With the verbal format, interested participants gave twice as many responses using banding only as standard participants.

Table 32: *Number of trials for verbal and bar chart formats, with banding at two-, three-, and four-levels, in which participants, with different levels of interest, gave reasons based only on banding information. The maximum number of trials for each standard or interested group is 16; the overall maximum, 32.*

Levels of banding		2	3	4
Verbal format:	standard	1	0	5
	interested	1	0	10
	overall	2	0	15
Bar chart format:	standard	4	0	9
	interested	2	0	8
	overall	6	0	17

We compared the number of reasons given in responses with labelling displaying seven nutrients and labelling displaying five nutrients (Table 33). Just as there had been shorter start times and response lengths, significantly fewer reasons were given when five nutrients were displayed compared to seven nutrients ($p < .01$). This difference was significant across all conditions.

Table 33: *Mean number of reasons given in responses when seven or five nutrients were listed (across participants with different levels of interest, seeing verbal and bar chart formats, with banding at two-, three-, and four-levels) in comparisons of two foods in Study 2. The range is shown in parentheses. (Note that the number of responses are shown to one decimal place here in order to clarify differences between conditions.)*

Number of nutrients displayed	7	5
Mean number of reasons in responses	3.1 (2–5)	2.7 (0–5)

We examined whether the different number of nutrients listed in banding displays affected the kind of reasons given in responses, as well as the number of reasons. Table 34 shows the reasons given with seven or five nutrients listed. The reasons are shown in rank order according to the frequency with which they were given. The number of trials in which each reason was given is shown (in parentheses) as a percentage of the total number of reasons given in each condition. Overall, there were similar patterns for seven and five nutrients, with fat and sugars being given priority across both conditions. However, protein level was given as a reason more frequently when participants saw banding for seven nutrients (where protein was the first nutrient in the banding panel) than when they saw five nutrients (where protein was not included in the banding panel). With seven nutrients, protein was mentioned more frequently than sodium or fibre, but this pattern was reversed with five nutrients.

106

Table 34: Reasons given in comparisons with seven or five nutrients listed in Study 2. The reasons are shown in rank order according to frequency. The number of trials on which each reason was given is shown (in parentheses) as a percentage of the total number of reasons given in each condition.

Number of nutrients displayed	7		5	
Reasons given in responses	Fat	(19%)	Fat	(20%)
	Sugars	(15%)	Sugars	(17%)
	Protein	(14%)	Sodium	(13%)
	Sat'd fat	(14%)	Sat'd fat	(12%)
	Sodium	(11%)	Fibre	(12%)
	Energy	(9%)	Energy	(10%)
	Carb'drate	(9%)	Protein	(8%)
	Fibre	(9%)	Carb'drate	(7%)

Participants made errors (for example, saying that the level of a nutrient was different from the level shown on the banding panels, or misreading the different levels of nutrients across two foods) for only 11 (4%) of the reasons they gave. The errors were distributed across the different conditions tested.

Participants rarely gave additional reasons for their judgements that could not be related to individual nutrients (reasons relating to their preconceptions about particular foods). Reasons based on preconceptions were given in six (3%) of the trials.

De-briefing

Table 35 shows participants' responses to questions about their preferences for verbal or bar chart formats. Standard participants preferred bar charts, but interested participants, were divided evenly between preferences for verbal and bar chart formats.

Table 35: Table 35 Number of standard and interested participants preferring different labelling formats used in Study 2.

Prefferred labelling format	verbal	bar chart	no difference
Standard	3	8	1
Interested	5	5	2

When participants who had received two-level or four-level banding were asked whether or not they preferred it to the three-level banding they had seen in Studies 1a and 1c, all participants who saw two-level banding said they preferred the three-level system. There was less agreement among participants who saw four-level banding schemes. In the standard group three said they would prefer three-level banding, and one said she would prefer a simple two-level banding of high and low (even though she had not used such a system in the comparison tasks). In the interested group, two preferred four-level banding used in Study 2 and two preferred the three-level banding. When interested participants gave these responses they commented on the trade-off between precision (for which they preferred four-level banding) and ease of use (for which they preferred three-level banding).

Participants were asked if they had noticed that in some cases seven nutrients had been listed in banding panels, and in other cases five. Three (out of twelve) standard participants and five (out of twelve) interested participants claimed to have noticed a difference. Table 36 shows that even though most participants had not noticed a difference, they preferred displays of seven, rather than five, nutrients.

107

Table 36: Participants in Study 2 preferring displays of seven or five nutrients.

Preferred number of nutrients	7	5
Preferred number of nutrients	7	5
Standard	10	2
Interested	9	3

7.4 Discussion of results in Study 2

Effects of labelling format (verbal versus bar chart)

Standard participants took less time to start their responses when they were presented with nutrition information in bar chart format compared to verbal format. There was no parallel effect for interested participants. Standard participants also expressed a preference for bar charts, which interested participants did not. However, using bar charts did not affect the length of time standard participants took to respond, nor the number of reasons they gave in their responses. It may be that the visual impact of bar charts helped speed up standard participants' pre-processing of information before they started their responses, or gave an impression of clarity which gave them confidence to start responding quickly, but they did not appear to confer any other information processing advantages. The impact of different kinds of graphic presentation of banding information was examined further in Study 3.

Effects of two-, three-, and four-level banding

The use of four, rather than two or three banding levels affected the performance of interested participants, who started responses more quickly with four-level banding, made shorter responses, and gave fewer reasons in those responses. Furthermore, with four-level banding, fewer of the reasons given in responses were based on numeric data; instead they were derived from the banding panels only (this was most obvious for interested participants, but was also a trend for standard participants).

It is likely that fewer participants used numeric information with four-level banding because it gave distinctions between foods that were detailed enough to make comparisons over a greater range of levels. (Participants using two-level banding, especially, commented on what appeared to them to be arbitrary cut-off points between the two banding levels.) However, the consequence of using banding information rather than numeric information was a tendency to cover fewer nutrients overall – just those that could be contrasted on the basis of their bandings. This may not mean that the decisions made with banding are any worse (or better) than decisions made with numeric information: they may be more focussed on significant differences between nutrient levels.

In de-briefing, all participants who saw two-level banding expressed a preference for three-level banding. However participants who saw four-level banding were divided about whether they would prefer three-level or four-level banding, recognising a trade-off between the relative simplicity of three levels and the greater precision of four. It was perhaps the tendency to recognise the benefits of four-level banding that made interested participants quicker to make the decision to use it, and more likely than standard participants to use it exclusively. This would be compatible with trends in Study 1a (where interested participants tended to gain more from banding than standard participants, possibly because they were more likely to recognise and exploit its advantages).

Similarly in Study 1c, interested participants seemed more likely than standard participants to use banding presented separately from numeric information exclusively, without making reference to numeric information.

Effects of presenting seven or five nutrients

The strongest effect in this study was the tendency for participants to take less time to start and complete their responses, and to give fewer responses, when banding information listed five rather than seven nutrients. Participants showed a strong preference for seven nutrients to be listed, on the grounds that they wanted to have the fullest information possible.

The analysis of the reasons given with the different displays showed that while, overall, participants' responses followed a generally similar pattern, participants were more likely to give priority to protein levels when they saw seven nutrients (where it was listed in the banding panel) than when they saw five nutrients (where it was not listed in the banding panel). Protein was one of the factors participants in the discussion group/ questionnaire study claimed influenced their decisions about foods to use or buy. So it was striking that many participants in this study only took protein into account when it was listed in the banding panel. It appears that the nutrients included, and the order of presentation of nutrients, in verbal or graphic displays (even when they are only a supplement to a fixed presentation of numeric information) influence the reasons underlying people's decisions about foods. In some cases, including or excluding particular nutrients may result in people taking account of factors that are not central to current concerns about nutrition and health.

Relating Study 2 to Studies 1a–c

We have shown similarities between this study and the previous studies in the performance of participants, particularly those interested in nutrition issues. Interested participants appeared to have a greater ability to capitalise on the different information formats used (see *Effects of two-, three, and four-level banding*).

More generally, this study complements the findings of Studies 1a–c by demonstrating that, although people want banding information, they see it as supplementary information, and will use numeric listings for detailed information. It is perhaps worth noting that participants in this study had, through their participation in Studies 1a and 1c, gained experience of using numeric information, and were also performing under less pressured conditions than many people making decisions during shopping. Consequently they may have been more likely to use numeric information than people in a real-life setting. Study 3, which was conducted with a group of participants who had not taken part in any previous studies, and so were relatively inexperienced at using both numeric and banding information, provided a good comparison with the present study.

In this study, banding was presented separately from numeric information. In Study 1c this type of format tended to cause errors, and was disliked, compared to banding presented alongside numeric information. However, very few errors were made here, possibly because participants had, by this stage, become used to seeing banding information separately.

8 Report of Study 3a

8.1 Aims

The aims of Study 3a were to compare people's use of:

- verbal banding (verbal format)
- verbal banding supplemented by an evaluative star system indicating the relative healthiness of the different levels of nutrients in a food (verbal plus stars format)
- an evaluative star system, indicating the relative healthiness of the different levels of nutrients, used without verbal banding (stars only format).

The different formats are shown in Figure 14. Banding or stars were always presented as a supplement to numeric nutrition information.

The different formats were all examined in comparisons of two foods, carried out by participants with different levels of interest in nutrition issues (standard and interested).

8.2 Method

Design

The design of the study is summarised in Table 37. We tested three groups of eight people, each seeing a different banding format: verbal format, verbal plus stars format, and stars only format. Within each group, half of the participants were standard, and half were interested, according to the screening questions used in recruitment. Each group saw the same set of materials, presented in the same order.

Table 37: Design of Study 3a. The number of participants in each treatment is shown in parentheses.

Banding format	Level of interest in nutrition issues
verbal (8)	standard (4)
	interested (4)
verbal plus stars (8)	standard (4)
	interested (4)
stars only (8)	standard (4)
	interested (4)

Materials

The labelling formats (shown in Figure 14) were based on the recommendations for the presentation of the 'Group 2' nutrients in the EC Nutrition Labelling Rules Directive (see notes on the typography and graphic design of the formats in Appendix 1, on pp. 131-142). The formats all consisted of a panel of numeric information (the same as the numeric format of Study 1a), with an additional panel giving banding in

Rice pudding

	per 100g	per serving 212g	
Energy	306	646	kJ
	72	152	kcal
Protein	3.7	7.8	g
Carbohydrate of which	12.4	26.3	g
Sugars	4.5	9.5	g
Fat of which	1.2	2.5	g
Saturated fats	0.8	1.5	g
Dietary fibre	0.2	0.4	g
Sodium	trace	trace	

Sugars	*high*
Fat	*low*
Saturated fats	*medium*
Dietary fibre	*low*
Sodium	*low*

Rice pudding

	per 100g	per serving 212g	
Energy	306	646	kJ
	72	152	kcal
Protein	3.7	7.8	g
Carbohydrate of which	12.4	26.3	g
Sugars	4.5	9.5	g
Fat of which	1.2	2.5	g
Saturated fats	0.8	1.5	g
Dietary fibre	0.2	0.4	g
Sodium	trace	trace	

The more stars shown for each nutrient, the healthier this food is for you

Sugars	high	
Fat	low	★★
Saturated fats	medium	★
Dietary fibre	low	
Sodium	low	★★

Rice pudding

	per 100g	per serving 212g	
Energy	306	646	kJ
	72	152	kcal
Protein	3.7	7.8	g
Carbohydrate of which	12.4	26.3	g
Sugars	4.5	9.5	g
Fat of which	1.2	2.5	g
Saturated fats	0.8	1.5	g
Dietary fibre	0.2	0.4	g
Sodium	trace	trace	

The more stars shown for each nutrient, the healthier this food is for you

Sugars	
Fat	★★
Saturated fats	★
Dietary fibre	
Sodium	★★

Figure 14: *Verbal banding, evaluative star system with verbal banding, and evaluative star system without verbal banding, tested in Study 3a.*

either verbal, verbal plus stars, or stars only format (details of the banding system are given in Appendix 4, on p. 146). Three levels of banding (high, medium and low) were used. The verbal format was the same as the format used for separate banding in Studies 1c and 2. In the verbal plus stars format, two stars were used to reinforce low bandings for sugars, fat, saturated fat, and sodium; and to reinforce high bandings for dietary fibre. One star was used to reinforce medium bandings for all the nutrients. No stars were used when sugars, fat, saturated fat, or sodium were high, or when fibre was low.

Note that we experimented with a direct mapping of nutrient levels to number of stars (so that, for example, fat would have three stars when low, two stars when medium and one star when high), but we rejected this alternative because of the ambiguity of giving stars to nutrient levels that would not be considered to be good for health.

It was impossible to give carbohydrate a star rating (since it includes sugars as well as other carbohydrates), and so it had to be excluded from the nutrient listing. Consequently we decided to use the five nutrient listing, tested in Study 2, which excludes protein, as well as carbohydrate, and focuses just on the nutrients central to current concerns about nutrition and health. The five nutrient listing was used in all conditions.

The labels in each comparison were presented side by side on flat sheets of paper. The name of each food was given above its label. Before the decision-making tasks, participants were shown the introductory summary used in Study 1a (the version without dietary reference values, see Appendix 3, on pp. 144-5).

111

Procedure

Participants were tested singly, in a quiet room. The procedure took about 15 minutes.

The background to the study and the procedure were outlined to the participants.

All participants were shown the introductory summary. They were asked to read it in their own time, and were told they could refer to it during the session if they wished.

The participants made eight comparisons of two foods. In each case participants were asked to say which food was the wiser choice for a healthy diet, and their responses were recorded in the same way as described in previous studies (see *Procedure* in Study 1a, section 4.2).

Participants

We tested 24 women, recruited using the same questionnaire as in the previous studies (see Appendix 2, on p. 143). The participants were screened to yield two groups of 12 participants: standard and interested. The participants had not taken part in any of the previous studies and so were relatively inexperienced at using the kind of nutrition information we were testing.

8.3 Results

The mean times for standard and interested participants to start responding with each banding format are shown in Table 38. Participants who saw the stars only format took significantly less time to start their responses than those who saw verbal plus stars format ($p < .01$). There was also a non-significant trend for participants to take less time to start their responses when they saw verbal plus stars format compared to verbal format. In all conditions, standard participants started their responses more quickly than interested participants (overall means, 18 and 22 seconds respectively, $p < .01$).

Table 38: Mean times (seconds) taken to start responses by participants with different levels of interest, seeing different banding formats in Study 3a. The range of times in each condition is shown in parentheses.

Banding format		verbal	verbal + stars	stars only
Interest level:	standard	22 (4–67)	19 (2–44)	13 (5–33)
	interested	27 (8–61)	22 (2–56)	19 (3–49)
	overall	25 (4–67)	21 (2–56)	16 (3–49)

The mean length of time participants took to complete responses with each banding format is shown in Table 39.

Table 39: Mean times (seconds) taken to complete responses by participants with different levels of interest, seeing different banding formats in Study 3a. The range of times in each condition is shown in parentheses.

Banding format		verbal	verbal + stars	stars only
Interest level:	standard	27 (9–70)	25 (9–50)	19 (3–37)
	interested	25 (9–50)	21 (8–47)	17 (3–39)
	overall	26 (9–70)	20 (9–47)	18 (3–38)

Participants took significantly less time to complete their responses when they saw the verbal plus stars format than when they saw the verbal format (p < .01). There was also a trend for participants to take less time to complete responses with the verbal plus stars format than with the stars only format, but this was not significant. There was no clear difference in the lengths of time taken by standard and interested participants.

The mean number of reasons given in responses with each banding format, and with each level of banding are shown in Table 40. There was a trend for participants to give fewer reasons with the verbal plus stars format than with the verbal format (p < .001), and a non-significant trend for them to give fewer reasons with stars only than with verbal plus stars. Interested participants gave significantly more reasons than standard participants (means 1.8 and 1.1 respectively, p < .001).

Table 40: Mean number of reasons given in responses by participants with different levels of interest, seeing different banding formats in Study 3a. The range is shown in parentheses.

Banding format		verbal	verbal + stars	stars only
Interest level:	standard	1.6 (0–4)	0.8 (0–2)	1.0 (0–3)
	interested	2.5 (0–5)	1.6 (0–3)	1.1 (0–2)
	overall	2.1 (0–5)	1.3 (0–3)	1.1 (0–3)

Table 41 shows the number of errors made in the reasons given with each of the labelling formats. (The errors are also shown as a percentage of the total number of reasons given in each condition). There were more errors with the stars only format than with either of the other formats, and more errors were made by standard participants than by interested participants.

Table 41: Mean number of errors made in reasons given in responses by participants with different levels of interest, seeing different banding formats in Study 3a. The errors are shown as a percentage of the total number of reasons given with each format in parentheses.

Banding format		verbal	verbal + stars	stars only
Interest level:	standard	0 (0%)	0 (0%)	12 (38%)
	interested	2 (2%)	2 (4%)	8 (23%)

We examined the errors and found that the four errors made by interested participants with verbal and verbal plus stars formats were the result of mis-readings of numeric information. All the other errors (all the errors for both standard and interested groups using the stars only format) were the result of participants misinterpreting the mapping of stars to nutrient levels. For example, when comparing two foods where the fat level of one was high, and the other low (see Figure 15), a participant might say that the food with the high fat level was a wiser choice because it had no star by fat, and so must have no fat; or a participant might say that the food with the low fat level (and so, two stars by fat) had more fat, and so the other food was a wiser choice. These errors were made for all the nutrients where the recommendation is for a low level (sugars, fat, saturated fat, sodium), and where low was marked with two stars. There were none of these kinds of errors for fibre,

113

where a high level was marked with two stars, and a low level with no stars. The mapping errors were distributed across participants. Four participants were inconsistent, interpreting the stars correctly in some comparisons, but not in others.

Rice pudding

	per 100g	per serving 212g	
Energy	306	646	kJ
	72	152	kcal
Protein	3.7	7.8	g
Carbohydrate of which	12.4	26.3	g
Sugars	4.5	9.5	g
Fat of which	1.2	2.5	g
Saturated fats	0.8	1.5	g
Dietary fibre	0.2	0.4	g
Sodium	trace	trace	

Gooseberry fool

	per 100g	per serving 114g	
Energy	667	774	kJ
	160	186	kcal
Protein	2.3	2.7	g
Carbohydrate of which	13.3	15.4	g
Sugars	13.3	15.4	g
Fat of which	11.1	12.9	g
Saturated fats	6.8	7.8	g
Dietary fibre	1.0	1.2	g
Sodium	trace	trace	

The more stars shown for each nutrient, the healthier this food is for you	
Sugars	
Fat	★★
Saturated fats	★
Dietary fibre	
Sodium	★★

The more stars shown for each nutrient, the healthier this food is for you	
Sugars	
Fat	
Saturated fats	
Dietary fibre	★
Sodium	★★

Figure 15: *In the evaluative star system without verbal banding, tested in Study 3a, mistakes were likely in comparisons of two foods where the fat level of one was low, and the other high (as for rice pudding and gooseberry fool, shown here). Participants tended to say that the food with the high fat level was a wiser choice because it had no star by fat, and so must have no fat; or participants might say that the food with the low fat level (and so, two stars by fat) had more fat, and so the other food was a wiser choice.*

Despite making errors in reasoning about individual nutrients, we thought participants still might be able to use the star system effectively to make good overall judgements about the relative healthiness of different foods. Consequently, we examined the responses participants gave, scoring errors when participants claimed that foods with fewer stars in the star-based systems were wiser choices for healthy eating than foods with more stars, (we used the same criteria for the verbal banding condition, even though participants were not using stars there). There were two comparisons where both foods had the same number of stars, and these were not scored. The total number of errors in overall judgements (maximum 24 for each sub-group of participants, 48 overall) are shown in Table 42. Significantly more errors were made overall when participants saw stars only than in either of the conditions where they saw verbal banding (p < .05 for both comparisons).

Table 42: Total number of errors made in responses (maximum 24 for each sub-group of participants, 48 overall) by participants with different levels of interest, seeing different banding formats in Study 3a.

Banding format		verbal	verbal + stars	stars only
Interest level:	standard	3	1	6
	interested	1	4	7
	overall	4	5	13

We isolated reasons given in responses that could not have been given on the basis of banding information, and where participants must have used the numeric information given in the label: for example, where participants mentioned calories, protein, or carbohydrates, which were not banded; or where they said there was a difference between levels of nutrient in different foods that fell within the same band in the banding panels. Table 43 shows the percentage of the responses given in each condition that could only have been given on the basis of numeric information. There was a trend for both groups of participants to use numeric information more often when they saw the verbal format than when they saw the stars only format. For interested participants, there was a trend to use numeric information more when they saw the verbal plus stars format, too.

Table 43: Percentage of reasons based on numeric information given by participants with different levels of interest, seeing different banding formats in Study 3a. The numbers represented by the percentages are shown in parentheses.

Banding format		verbal	verbal + stars	stars only
Interest level:	standard	52% (26)	22% (6)	22% (7)
	interested	70% (56)	75% (40)	53% (19)

Analysis of the reasons given showed that there were some comparisons where participants did not look at individual items in either the banding or numeric information, but simply looked at the banding, and made a judgement on the basis of the total number of stars shown. The number of trials (maximum 32) where participants used this strategy are shown in Table 44. It was used in both verbal plus stars and stars only format, although more often by standard participants. (Note that it would have been possible for participants to use a similar strategy with verbal formats, by counting the number of 'lows' for nutrients other than fibre.)

Table 44: Number of comparisons (maximum 32) where participants gave reasons based on the number of stars displayed in banding (for participants with different levels of interest, seeing different banding formats in Study 3a).

Banding format		verbal	verbal + stars	stars only
Interest level:	standard	0	7	7
	interested	0	1	4

8.4 Discussion of findings in Study 3a

The visual impact of star systems

Using stars tended to reduce the time participants took to start their responses, and to reduce the length of time taken to respond, and the number of reasons given in responses. These effects were similar to the effect of bar charts, observed in some conditions in Study 2. However the effect of the stars was more marked than the effect of bar charts. This difference may have arisen because participants in this study were relatively inexperienced at using numeric and verbal banding information compared to participants in Study 2, and so may have been more easily diverted from examining numeric information by a graphic representation. Alternatively it may have been that using stars as evaluative tokens gives a signal that is qualitatively different from bar charts, and has a greater impact (the evaluative star systems used here might be thought of as having a stronger semantic force than the relatively neutral bar charts used in Study 2).

As with the bar charts, the effect of the stars appeared to be greatest for standard participants. However, whereas the bar charts simply reduced the start time for participants' responses, stars not only reduced response times, but also affected the quality of the responses. They significantly reduced the number of reasons given in responses, and, when used without verbal banding, increased the number of errors made in reasons given in responses (particularly for standard participants), and decreased the overall accuracy of the responses.

The errors could be traced directly to the way in which the stars mapped on to banding information. Participants did not have difficulties in interpreting the significance of the nutrient levels (high, medium, and low) in verbal banding, or in using numeric information. In both those cases they had to draw on their understanding of nutrition issues in order to interpret the banding or numeric information correctly, just as participants had done when using bar charts in Study 2. In contrast, the evaluative star system indicated the relative healthiness of nutrient levels, so that participants did not have to draw on their own understanding in comparison tasks. This evaluative system proved a distraction.

It could be that in a different kind of task, such as making simple choices between foods without having to give reasons for those choices, star systems would provide better support for decisions than numeric information or verbal banding. People would only have to count up the number of stars for each food, and would be able to make quick decisions on the basis of a strategy of 'the more stars the better'. However, in the kind of decision-making task we examined here, where people considered the levels of individual nutrients in the process of making choices between foods, stars led to confusions. If the goal of nutrition information is to help consumers make good decisions across a range of tasks, star systems might be seen as a less than satisfactory option. A compromise would be to use both stars and verbal banding. Our results showed that this format was less prone to error than stars only.

These findings gave support to the finding of Study 2, that graphic representations of banding influenced decision-making. In Study 3b we examined an alternative graphic format, but one that, like bar charts, mapped directly on to banding levels, without the evaluative aspect of the star system used here.

Performance of standard and interested participants

In Studies 1a–c and 2 we observed that interested participants tended to be more likely than standard participants to adapt their information gathering strategies to the format with which they were presented, and, particularly, to make more use of non-numeric information than standard participants. In this study the pattern appears to be reversed since interested participants generally took longer to use all formats than standard participants, and tended to make more use of numeric information than standard participants. However, this may be part of a more general ability to exploit different sources of information in different circumstances.

In the earlier studies, standard participants showed greater rigidity than interested participants, by sticking to the numeric information with which they had become familiar. Here, standard participants who were not familiar with numeric information showed greater rigidity by sticking to banding information. Aware participants showed greater flexibility by moving between different information sources: they not only consulted numeric information more often, but also used the strategy of looking at the total number of stars shown. The use of different sources of information may explain why interested participants took longer to start and complete their responses, but they also gave more reasons for their decisions, and were less prone to errors than standard participants. It is ironic that the stars only system, which might have been expected to help standard participants, since it avoided the need for any evaluation of banding information, caused them particular problems.

9 | Report of Study 3b

9.1 Aims

The aims of Study 3b were to compare people's use of:

- verbal banding (verbal format)
- combined verbal banding and bar charts (bar chart format, as tested in Study 2)
- combined verbal banding and boxes shaded with different tints, to indicate different levels of nutrient (shaded box format).

The different formats are shown in Figure 16. Banding was always presented as a supplement to numeric nutrition information.

The different formats were all examined in comparisons of two foods, carried out by participants with different levels of interest in nutrition issues (standard and interested).

9.2 Method

Design

The design of the study is summarised in Table 45. We tested three groups of eight people: one group saw banding in a verbal format, one saw a bar chart format, and one saw a shaded box format. (See notes on the allocation of participants to conditions under *Participants*, below).

Within each group, half of the participants were standard, and half interested, according to the screening questions used in recruitment to determine participants' level of interest in nutrition issues. Each group saw the same set of materials, in the same order.

Table 45: *Design of Study 3b. The number of participants in each condition is shown in parentheses.*

Banding format	Level of interest in nutrition issues
verbal (8)	standard (4)
	interested (4)
bar chart (8)	standard (4)
	interested (4)
shaded boxes (8)	standard (4)
	interested (4)

Materials

The labelling formats (shown in Figure 16) were based on the recommendations for the presentation of the 'Group 2' nutrients in the EC Nutrition Labelling Rules Directive (see notes on the typography and graphic design of the formats in Appendix 1). The formats all consisted of a panel of numeric information, with an additional panel beneath,

Rice pudding

	per 100g	per serving 212g	
Energy	306	646	kJ
	72	152	kcal
Protein	3.7	7.8	g
Carbohydrate of which	12.4	26.3	g
Sugars	4.5	9.5	g
Fat of which	1.2	2.5	g
Saturated fats	0.8	1.5	g
Dietary fibre	0.2	0.4	g
Sodium	trace	trace	

Sugars	*high*
Fat	*low*
Saturated fats	*medium*
Dietary fibre	*low*
Sodium	*low*

Rice pudding

	per 100g	per serving 212g	
Energy	306	646	kJ
	72	152	kcal
Protein	3.7	7.8	g
Carbohydrate of which	12.4	26.3	g
Sugars	4.5	9.5	g
Fat of which	1.2	2.5	g
Saturated fats	0.8	1.5	g
Dietary fibre	0.2	0.4	g
Sodium	trace	trace	

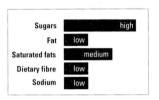

Rice pudding

	per 100g	per serving 212g	
Energy	306	646	kJ
	72	152	kcal
Protein	3.7	7.8	g
Carbohydrate of which	12.4	26.3	g
Sugars	4.5	9.5	g
Fat of which	1.2	2.5	g
Saturated fats	0.8	1.5	g
Dietary fibre	0.2	0.4	g
Sodium	trace	trace	

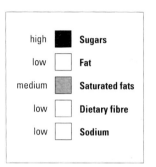

Figure 16: *Verbal banding, bar charts, and shaded boxes, tested in Study 3b.*

giving banding. The verbal format was the same as the format used in Studies 1c, 2, and 3a. The bar chart format was the same as the format used in Study 2. In the shaded box format the nutrient names were placed to the right of the words and boxes showing banding levels (see Appendix 1, p. 131-142, for details of the design of the formats).

Three levels of banding (high, medium, low) were used. In the shaded box format, a black box was used to indicate a high level of a nutrient, a grey tinted box to indicate a medium level, and a white box to indicate a low level. So there was a direct mapping from banding words (high, medium, low) to their graphic representation by shaded boxes. This contrasted with Study 1a, where the mapping of stars was from the benefits for health of the nutrient level to the number of stars.

The shaded box system confined us to three banding levels, since we felt that more than three levels of shading (black, grey and white) might not be discriminated easily by consumers, and the difficulty of discrimination could be exacerbated by variation in reproduction across different printed food packages (see Appendix 1). For comparability with Study 3a we continued to use the five nutrient listing, excluding carbohydrate and protein. However, the shaded box format could have been used to represent carbohydrate, since it was a direct mapping, rather than an evaluation of banding levels (see *Materials* section in Study 3a).

The labels in each comparison were presented side by side on flat sheets of paper. The name of each food was given above its label, as in the previous paper-based studies.

119

Procedure

Participants were tested singly, in a quiet room. The procedure followed directly from Study 3a, and took about 7 minutes.

The participants made four comparative judgements about foods, using nutrition information printed on sheets of paper. All the judgements were comparisons of two foods. In each case participants were asked to say which food was the wiser choice for a healthy diet, and their responses were recorded in the same way as described in the previous studies (see detailed description in *Procedure* for Study 1a).

Participants

The participants were the same as in Study 3a. In order to control practice effects, we made sure that participants who had seen verbal banding in 3a did not see verbal banding again in this study. Consequently, all the participants had experience of using a banding system, but not the same system as they saw in this study.

9.3 Results

The mean times for standard and interested participants to start responding with each banding format are shown in Table 46. Although there was a trend for participants seeing shaded boxes to take longer to start their responses than participants seeing verbal or bar chart formats, this was not significant. Overall, interested participants took longer to start their responses than standard participants (means 19 and 13 respectively, $p < .05$).

Table 46: *Mean times (seconds) taken to start responses by participants with different levels of interest, seeing different banding formats in Study 3b. The range of times in each condition is shown in parentheses.*

Banding format		verbal	bar charts	shaded boxes
Interest level:	standard	11 (4–44)	9 (5–17)	18 (3–54)
	interested	20 (18–60)	18 (5–49)	20 (5–50)
	overall	16 (4–60)	14 (5–49)	19 (3–54)

The mean length of time participants took to complete responses with each banding format is shown in Table 47. Responses were significantly shorter when participants saw verbal formats than when they saw bar chart formats ($p < .001$). There was no significant difference in response length for bar chart and shaded boxes. The difference between response length for verbal and bar chart formats was mainly due to the interested participants. However, there was no significant difference overall in the response lengths of standard and interested participants (means 17 and 18 seconds, respectively).

Table 47: *Mean times (seconds) taken to complete responses by participants with different levels of interest, seeing different banding formats in Study 3b. The range of times in each condition is shown in parentheses.*

Banding format		verbal	bar charts	shaded boxes
Interest level:	standard	13 (7–30)	18 (6–37)	20 (7–74)
	interested	10 (3–27)	22 (9–45)	21 (7–60)
	overall	12 (3–30)	20 (6–45)	21 (7–74)

The mean number of reasons given in responses with each banding format, and with each level of interest are shown in Table 48. Participants gave significantly more reasons when they saw bar charts than when they saw the verbal format ($p < .05$); and they gave significantly more reasons when they saw bar charts than when they saw shaded boxes ($p < .01$). Note that difference in number of reasons given with verbal and bar chart formats, like the difference in response lengths, is particularly due to the responses of interested participants. Overall, however, there was no significant difference between the number of reasons given by standard and interested participants (means 1.4 and 1.5, respectively).

Table 48: Mean number of reasons given in responses by participants with different levels of interest, seeing different banding formats in Study 3b. The range is shown in parentheses. (The number of reasons are shown to one decimal place in order to clarify the differences between conditions.)

Banding format		verbal	bar charts	shaded boxes
Interest level:	standard	1.7 (1–3)	1.6 (0–3)	0.8 (0–2)
	interested	1.1 (0–2)	2.1 (1–4)	1.4 (0–3)
	overall	1.4 (0–3)	1.9 (0–4)	1.2 (0–3)

Only four errors were made in the reasons given, and these were all made by standard participants seeing the shaded box format. Two of those errors were mapping errors made by a participant who had seen the verbal plus stars format in Study 3a. She judged foods with high levels of nutrients such as sugars, fat, saturated fats and sodium as wise choices for a healthy diet because they had more black boxes in their banding displays than the foods with which they were being compared.

We analysed the reasons given to isolate reasons that could not have been given on the basis of banding information, and where participants must have used the numeric information given in the label. Table 49 shows the percentage of the responses given in each condition that could only have been given on the basis of numeric information.

Table 49: Percentage of reasons based on numeric information given by participants with different levels of interest, seeing different banding formats in Study 3b. The numbers represented by the percentages are shown in parentheses.

Banding format		verbal	bar charts	shaded boxes
Interest level:	standard	0% (0)	32% (8)	38% (5)
	interested	0% (0)	47% (16)	45% (10)

Participants were more likely to resort to numeric information when they saw bar chart or shaded box formats than when they saw verbal formats. Overall, interested participants tended to use numeric information more often than standard participants.

Analysis of the reasons given showed that there were some comparisons where participants did not look at individual items in either the banding or numeric information, but simply looked at the banding, and made a judgement on the basis of the total number of boxes shown (a strategy they would have learned when seeing the star banding systems used in Study 3a). Participants only used this strategy when they saw shaded boxes: standard participants used the strategy in six (out of a maximum of 16) comparisons, and interested participants in one (out of a maximum of 16) comparisons. Note that on two of the six occasions where a standard participant used this strategy, she used it incorrectly (see discussion of errors, above).

9.4 Discussion of findings in Study 3b

In contrast to Study 3a, where using stars led to shorter responses, the graphic representations of banding used here (bar charts and shaded boxes) led to longer responses and, for shaded boxes, less detailed responses. This may have been the consequence of a transfer effect: participants were used to a graphic representation that did not map directly on to banding, and now had to learn how to use representations that did map on to banding. Participants were more successful in making the transfer with bar charts. The particular difficulty with the shaded boxes may have been their ambiguity: they may have been perceived as similar to the star system, with a darker box indicating 'better for health' or, conversely, a lighter box indicating 'better for health'; or they could have been perceived (correctly, in this case) as a direct mapping system, with a darker box meaning 'more of a nutrient'. The mapping errors made by one participant seemed to confirm that she was attempting to use the shaded boxes in the same way as the star system she had seen previously.

Although these transfer effects make it difficult to draw conclusions about the systems tested, the confusion of the shaded box system in this study and the star system in Study 3a illustrates the difficulty of having a range of systems in operations simultaneously (for example, on supermarket shelves). Given the number of errors due to misinterpretation of the evaluations given in the star system, despite the (slightly) higher number of reasons given in responses in this study, it is reasonable to suggest that graphic representations of banding should be limited to one kind of mapping only, and that the safest option may be a direct, rather than an evaluative, mapping system.

10 Report of Study 4

10.1 Aims

The aim of Study 4 was to examine three nutrition label formats in which the nutrient levels were related to dietary reference values (DRVs):

- a format giving DRVs alongside the standard listing of numeric quantities for each nutrient (numeric DRV format)
- a format showing the percentage of the DRV for each nutrient in a serving of the food, alongside the standard listing of numeric quantities (percentage DRV format)
- a format with the percentage of the DRVs, given as a percentage and represented graphically, in a separate display from the standard listing of numeric quantities (graphic DRV format).

The three formats are shown in Figure 17, on p. 124.

The formats were examined in evaluations of single foods and comparisons of two foods. There were two groups of participants in the study (standard and interested), and level of interest was included in the analysis.

10.2 Method

Design

The design of the study is summarised in Table 50. We tested 24 participants. The participants were divided into two groups (standard, and interested), according to the screening questions used in recruitment to determine level of interest in nutrition issues. Each participant saw all three DRV formats. There were three sets of materials (I, J, and K). The allocation of materials to formats was varied systematically within each group of participants, and the order of presentation of the formats was rotated across participants.

Table 50: *Design of Study 4. The number of participants in each treatment is shown in parentheses.*

Interest level	Allocation of materials (sets I, J, K) to DRV formats
standard (12)	I numeric; J percentage ; K graphic (4) K numeric; I percentage ; J graphic (4) J numeric; K percentage ; I graphic (4)
interested (12)	I numeric; J percentage ; K graphic (4) K numeric; I percentage ; J graphic (4) J numeric; K percentage ; I graphic (4)

The main study was followed by a separate de-briefing study in which three groups of eight participants (four interested and four standard) described their use of different nutrition information formats presented on food packages.

Chicken korma

	per 100g	per serving 340g	recommended daily intake*	
Energy	921	3130	9450	kJ
	221	750	2400	kcal
Protein	12.5	45.0	75	g
Carbohydrate of which	11.0	37.0	345	g
Sugars	6.9	23.5	72	g
Fat of which	14.4	49.0	80	g
Saturated fats	7.3	24.8	27	g
Dietary fibre	4.5	15.0	30	g
Sodium	0.4	1.3	2	g

*recommended daily intake based on a 2400 kcal per day diet

Chicken korma

	per 100g	per serving 340g		percentage of recommended daily intake* (per serving)
Energy	921	3130	kJ	
	221	750	kcal	
Protein	12.5	45.0	g	60%
Carbohydrate of which	11.0	37.0	g	11%
Sugars	6.9	23.5	g	33%
Fat of which	14.4	49.0	g	61%
Saturated fats	7.3	24.8	g	92%
Dietary fibre	4.5	15.0	g	50%
Sodium	0.4	1.3	g	65%

*recommended daily intake based on a 2400 kcal per day diet

Chicken korma

	per 100g	per serving 340g	
Energy	921	3130	kJ
	221	750	kcal
Protein	12.5	45.0	g
Carbohydrate of which	11.0	37.0	g
Sugars	6.9	23.5	g
Fat of which	14.4	49.0	g
Saturated fats	7.3	24.8	g
Dietary fibre	4.5	15.0	g
Sodium	0.4	1.3	g

	recommended daily intake*	
Protein	60%	● ● ● ● ● ● ○ ○ ○ ○
Carbohydrate	11%	● ○ ○ ○ ○ ○ ○ ○ ○ ○
Sugars	33%	● ● ● ○ ○ ○ ○ ○ ○ ○
Fat	61%	● ● ● ● ● ● ○ ○ ○ ○
Saturated fats	92%	● ● ● ● ● ● ● ● ● ○
Dietary fibre	50%	● ● ● ● ● ○ ○ ○ ○ ○
Sodium	65%	● ● ● ● ● ● ● ○ ○ ○

* the black circles show the percentage of recommended daily intake per serving (based on a 2400 kcal per day diet)

Figure 17: Numeric dietary reference values, percentage dietary reference values, and graphic dietary reference values, tested in Study 4.

Materials

The labelling formats (shown in Figure 17) were based on the recommendations for the presentation of the 'Group 2' nutrients in the EC Nutrition Labelling Rules Directive (see notes on the typography and graphic design of the formats in Appendix 1, pp. 131-142). The formats all consisted of a panel of numeric information. In the numeric DRV format, DRVs were added in a tinted panel, alongside the numeric information; the percentage DRVs were also placed alongside the numeric formats. In the graphic DRV format, a separate panel was added beneath the numeric information panel, listing the nutrients with a percentage DRV, and representing the percentage on a scale of ten circle units. Note that on the labelling we described DRV as 'recommended daily intake' to ensure that participants understood what it represented (see Appendix 4, p. 146, for details of the DRV system used).

The labels in each comparison were presented either singly or side by side on flat sheets of paper. The name of each food was given above its label, as in the previous paper-based studies.

In the de-briefing study the three different types of DRV formats were applied to food packages, along with a separate panel giving verbal banding, the same as the banding panels showing seven nutrients tested in Studies 1c and 2. The banding was positioned beneath the DRV display. This information was substituted for any existing nutrition labelling on the food packages used. As far as possible the colour schemes used on the packages were reproduced on the substitute labels.

Procedure

Participants were tested singly, in a quiet room. The procedure followed from Study 3b after short break, and took about 15 minutes.

The participants made two evaluations of single foods, with each DRV format (six judgements in all). They then made two comparisons of two foods, with each DRV format (six comparisons). In the evaluations of single foods, participants were asked to say whether the foods were a wise choice for a healthy diet; in the comparisons they were asked to say which of the two foods was the wiser choice for a healthy diet, and their responses were recorded in the same way as described in the previous studies (see detailed description in *Procedure* for Study 1a, section 4.2).

In the de-briefing task, we presented participants with food packages, two singly, followed by two sets of two packages. We asked participants to make judgements of the healthiness (or relative healthiness) of the foods, and to say which information on the packages they were using as they made their judgements. Before they started the task we pointed out the different sources of information on the packages, showing the numeric listings, banding, and DRVs (with different formats, according to which of the three groups participants were in). After participants had carried out the de-briefing task we asked them directly about their preferences for the different formats they had seen.

Participants

The participants were the same as in Studies 3a and b.

10.3 Results

Judgements of single foods

The mean times for standard and interested participants to start responding with each DRV format are shown in Table 51. Overall there were no significant differences across the conditions. However, when we analysed the results for standard participants separately, we found they took significantly less time to start their responses with the graphic DRV format than with the numeric DRV format (p < .01). While there was a trend for their start times to be longer with percentage formats than with numeric formats, this difference was not significant. Overall, the start times were significantly shorter for standard participants than for interested participants (means 15 and 23 seconds respectively, p < .01).

Table 51: Mean times (seconds) taken to start responses by participants with different levels of interest, seeing different DRV formats in judgements of single foods in Study 4. The range of times in each condition is shown in parentheses.

DRV format		numeric	percentage	graphic
Interest level:	standard	17 (12–26)	23 (6–50)	8 (4–16)
	interested	22 (11–38)	23 (7–52)	24 (6–40)
	overall	19 (11–38)	23 (6–52)	17 (4–40)

The mean length of time participants took to complete responses with each banding format is shown in Table 52. Participants gave significantly shorter responses when they saw graphic DRV formats than when they saw percentage DRV formats (p < .01), and this effect was greater for standard than for interested participants. There was a trend for participants to give shorter responses with percentage DRVs than with numeric DRVs, but this was not significant. There was no overall difference in the response lengths of standard and interested participants (means 22 and 25 seconds respectively).

Table 52: Mean times (seconds) taken to complete responses by participants with different levels of interest, seeing different DRV formats in judgements of single foods in Study 4. The range of times in each condition is shown in parentheses.

DRV format		numeric	percentage	graphic
Interest level:	standard	28 (11–44)	25 (9–35)	14 (5–25)
	interested	29 (22–44)	27 (18–36)	20 (2–36)
	overall	29 (11-44)	26 (9-36)	18 (2-36)

The mean number of reasons given in responses with each format are shown in Table 53. Although there was a trend for participants to give more reasons when they saw numeric DRV formats than when they saw percentage DRV formats, this was not significant. However participants gave significantly more responses when they saw graphic DRV formats than when they saw percentage DRV formats.

Table 53: Mean number of reasons given in responses by participants with different levels of interest, seeing different DRV formats in judgements of single foods in Study 4. The range in each condition is shown in parentheses.

DRV format		numeric	percentage	graphic
Interest level:	standard	2.1 (0–5)	1.6 (0–4)	2.3 (1–3)
	interested	2.8 (0–5)	1.6 (0–3)	2.4 (1–3)
	overall	2.5 (0–5)	1.6 (0–4)	2.4 (1–3)

The nutrient levels mentioned in responses varied across the different conditions. However, because of the relatively low number of reasons given (compared, for example, to judgements of single foods in Study 1a) there were no clear trends in the changes between conditions.

The provision of DRVs meant that participants did not have to interpret the quantities of nutrients in the same way as they had done when examining single foods with numeric information only (in Study 1a). However participants still made some errors of interpretation (see Table 54). Most errors were made when participants saw numeric DRV formats. Overall standard participants made more errors than interested participants.

Table 54: *Percentage errors made by participants with different levels of interest, seeing different DRV formats in judgements of single foods in Study 4. The number represented by each percentage is shown in parentheses.*

DRV format		numeric	percentage	graphic
Interest level:	standard	16% (8)	12% (5)	11% (6)
	interested	9% (6)	5% (2)	2% (1)

As with judgements of single foods in Study 1a, participants sometimes gave preconceived reasons in their evaluations – in 15% (21) of the trials, distributed across the different conditions.

Comparisons of two foods

The mean times for standard and interested participants to start responding with each DRV format are shown in Table 55. Participants took significantly less time to start their responses when they saw graphic DRV formats than when they saw numeric DRV formats ($p < .05$). There was a trend for participants (particularly standard participants) to take less time to start their responses when they saw numeric DRV formats rather than percentage DRV formats.

Table 55: *Mean times (seconds) taken to start responses by participants with different levels of interest, seeing different DRV formats in comparisons of two foods in Study 4. The range of times in each condition is shown in parentheses.*

DRV format		numeric	percentage	graphic
Interest level:	standard	15 (9–40)	20 (7–51)	13 (7–26)
	interested	31 (8–50)	31 (13–57)	20 (4–54)
	overall	24 (8–50)	26 (7–57)	17 (4–54)

The mean length of time participants took to complete responses with each DRV format is shown in Table 56. There were no significant differences between the times taken to complete responses across conditions.

Table 56: *Mean times (seconds) taken to complete responses by participants with different levels of interest, seeing different DRV formats in Study 4. The range of times in each condition is shown in parentheses.*

DRV format		numeric	percentage	graphic
Interest level:	standard	22 (12–52)	27 (20–34)	23 (11–43)
	interested	26 (15–34)	28 (11–36)	24 (13–35)
	overall	25 (12–52)	28 (11–36)	24 (11–43)

The mean number of reasons given in responses with each DRV format are shown in Table 57. There were no significant differences in the numbers of reasons given in each condition. There was a trend for interested participants to give more reasons than standard participants (means 2.1 and 2.5 respectively), but this was not significant.

Table 57: Mean number of reasons given in responses by participants with different levels of interest, seeing different DRV formats in comparisons in Study 4. The range in each condition is shown in parentheses.

DRV format		numeric	percentage	graphic
Interest level:	standard	2.0 (1–3)	2.1 (1–3)	2.3 (1–5)
	interested	2.2 (0–5)	2.9 (2–5)	2.5 (1–4)
	overall	2.1 (0–5)	2.6 (1–5)	2.4 (1–5)

There were very few errors: two by standard participants using numeric DRV formats, and one by an interested participant using a graphic DRV format.

De-briefing task

Table 58 shows the number of trials (maximum eight) in each condition when participants said they were using particular sources of information, for both judgements of single foods and comparisons. Sometimes participants mentioned more than one source of information, but this information has not been included in Table 58. On some occasions participants did not use nutrition information, relying either on their own knowledge of the foods, or on other information on the packaging (in 10% of trials for single foods, and 6% of trials for two foods). Therefore the number of consultations recorded is less than the total number of trials (eight) in some conditions.

Table 58: Sources of information (DRV, banding, or numeric listing of nutrient quantities) used by participants judging single foods and comparing foods in de-briefing task in Study 4. The number of participants using each source of information (maximum eight) is shown. On some trials participants did not use nutrition information, so the number of consultations recorded is less than the total number of trials (eight) in some conditions.

	Judgements of single foods	Comparisons of two foods
Standard participants		
DRV format		
Numeric	2 DRV 5 banding	4 DRV 3 banding
Percentage	3 DRV 5 banding	3 DRV 4 banding
Graphic	5 DRV 2 banding	5 DRV 2 numeric
Interested participants		
DRV format		
Numeric	6 DRV	8 DRV
Percentage	8 DRV	7 DRV 1 banding
Graphic	6 DRV 1 banding	4 DRV 4 numeric

The responses suggested that standard participants were less likely to use the numerical or percentage DRV information than interested participants. In judgements of single foods, standard participants tended to use banding more often than DRVs presented numerically or as percentages, but they used DRVs presented in graphic formats in preference to banding. In contrast, interested participants used banding only as a back-up to DRVs presented numerically or as percentages, although when DRVs were presented as graphic formats they tended to use them in preference to other formats. In comparisons of two foods, standard participants still tended to use banding, except when DRVs were presented graphically. Interested participants rarely used banding, and did not use the graphic formats for DRVs as often as standard participants.

When we asked participants directly about their preferences, five out of twelve standard participants thought DRVs were more helpful than banding information, two thought banding was more useful, and five thought there was no difference between the two. For the interested participants, seven thought DRVs were more useful than banding, one thought banding more useful than DRVs and four thought there was no difference between the two. Two of the interested participants mentioned that there was sometimes a trade-off between speed and detail of information in choosing between the two. They referred, in particular, to two foods in a comparison, where both were banded as high in salt. In one of the foods, the salt level exceeded the DRV (it was three times the DRV) but in the other, although high, it was still less than the DRV. This contrast was obscured by the banding information. (Similar observations about the trade-off between speed and accuracy of detail were made by interested participants when they were using different levels of banding in Study 2, section 7.3.)

10.4 Discussion of Study 4

Representing dietary reference values

The most striking result of this study was the speed with which standard participants started their responses, both in judgements of single foods and comparisons of foods, when they saw graphic representations of DRVs. Furthermore in judgements of single foods, standard participants took less time to complete their responses with graphic DRV formats than with the other formats, without the same level of consequences for accuracy observed in Studies 3a and 3b. The advantage of the graphic DRV format was that it mapped quantities of nutrients directly, unlike the star system used in 3a, and, possibly, gave a less ambiguous representation of quantity, or level, than the shaded box format used in Study 3b. This result gives further support for graphic systems which give direct mappings, and which people can interpret using their existing understanding of nutrition issues, rather than evaluative graphic systems.

The relatively poor performance of all participants with percentage DRV formats was surprising. It may have been that because these participants were used to seeing separate displays of banding information (in Studies 3a and 3b), and often used them as their first, or only, source of information, they found it strange to use a composite display of numeric and supplementary information.

We might have expected the percentage DRV format to be easier to use than the numeric format, since it eliminated the need for participants to make comparisons of the nutrient levels to reference values. There was a trend for fewer errors to be made with percentage DRVs than with numeric DRVs, and interested participants made more effective comparisons of two foods using percentages than numeric DRVs. Otherwise it appeared as if participants were distracted by the use of percentages (one participant in the standard group volunteered that she found percentages difficult to use), but percentages were also used in the graphic DRV formats. Also we would not have expected the interested participants to have been distracted by using percentages, and, indeed, their subsequent use of percentage DRV formats in the de-briefing task suggested that generally, they were not.

Judgements made using numeric DRVs tended to be slower, and in some cases, less detailed than judgements made using graphic DRV formats. However, they were, at least, less prone to error than judgements made using numeric information on its own: the errors in the judgements of single foods here were 16% compared to 48% in Study 1a. Providing simple numeric reference values reduced the number of errors made through misinterpreting the quantities of nutrients in foods.

Relating Study 4 to previous studies

In this task, as in Studies 3a and b, interested participants tended to take longer to respond than standard participants; they also tended to give more, correct reasons than standard participants. The de-briefing task suggested that interested participants were, possibly, more selective in the information sources they used: they tended to used numeric or percentage DRV formats more than standard participants, particularly in the comparisons tasks.

Throughout this study participants gave more reasons in their responses than in Studies 3a and b. However we should be cautious about concluding from this that giving DRVs necessarily encourages people to be more analytical in their decisions about foods. One factor influencing the number of reasons given here may have been the use of displays for seven nutrients, rather than the five nutrient displays used in Studies 3a and b. This meant judgements about protein and carbohydrate could be made without consulting numeric information, and so increased the range of responses that could be given on the basis of just one source of information.

Although participants gave more reasons in their responses here than in Studies 3a and b, they tended to give fewer reasons than participants in Studies 1a–c and 2. This is likely to have been a result of the different experience these participants had had in using nutrition information. Participants in the earlier studies had used numeric and verbal banding information before seeing any graphic representations of nutrition information. Their strategy, and in particular the strategy of the standard participants, was to give priority to numeric information. This contrasted with the participants here, whose experience of using nutrition information had been focussed on graphic representations. The development of distinct patterns of using nutrition information within the short span of a testing session suggests that education in the use of a particular format of nutrition information may not be difficult in itself, but that education to help people use a range of different kinds of nutrition information formats might be considerably more difficult.

Appendix 1

Notes on the typography and graphic design of the labels tested

Note that the formats shown here were drafted and reproduced for the label sizes used in testing, where black type was displayed on a plain white or light-coloured background. The typeface, type sizes, weights etc. might need adjustment to suit changes in output resolution, production technology, size or colour.

Numeric format

We aimed to design a format that would make it easy for people to scan down the list of nutrients to any particular nutrient, and to scan across from that nutrient name to the related quantities.

Nutrient names

The nutrient names were ranged left (1). Because the names varied in length, ranging left meant that short names (such as 'Fat') were further away from the related figures than long names (such as 'Carbohydrate'). We experimented with ranging the names right (2) in order to keep names and figures consistently close together, but found that the shorter names were difficult to pick out, and the inclusion relationships ('Carbohydrate, of which Sugars' and 'Fats, of which Saturated fats') were obscured. We used horizontal rules to improve scanning across from names to two columns of figures.

The horizontal rules had a second function, of grouping related information. They were used selectively to make clear the equivalence of kilocalories and kilojoules, and the inclusion relationships between carbohydrate and sugars, and fats and saturated fats (3). We emphasised the inclusion relationships by using the words 'of which', commonly used in nutrition labelling, and compulsory according to the EC Nutrition Labelling Rules Directive. We also reduced slightly the size of the type for the included nutrients (sugars and saturated fats). We experimented with indenting (4) and bracketting (5) the names of the included nutrients, as a further cue to inclusion. But we rejected these alternatives, since the included nutrients are significant for health, and we did not want to imply that they were of subsidiary importance, nor to make them difficult to find as people scanned down the nutrient list.

The nutrient names were all presented in Univers bold condensed. The cues to inclusion 'of which' were set in plain Univers, at a small size, in order to minimise any disruption to the vertical scanning of the nutrient list. We rejected setting these words in italic, because we felt it would attract attention to them (6).

Nutrient quantities

The nutrient quantities were listed in two separate columns (per 100g and per serving). They were ranged left to facilitate scanning (i.e. the reader could expect the quantity information to be shown at a predictable point on the label). Often quantities on nutrition labels are aligned vertically about their decimal points (7), sometimes with vertical rules separating the columns of nutrient names and quantities, and the columns of listings per 100g and per serving (8). We rejected these kinds

Numeric format

1

	per 100g	per serving 330g	
Energy	659	2175	kJ
	157	518	kcal
Protein	7.8	25.7	g
Carbohydrate	10.2	33.7	g
of which **Sugars**	1.1	3.6	g
Fat	10.0	33.0	g
of which **Saturated fats**	4.7	15.5	g
Dietary fibre	3.0	9.9	g
Sodium	0.3	1.1	g

2

	per 100g	per serving 330g	
Energy	659	2175	kJ
	157	518	kcal
Protein	7.8	25.7	g
Carbohydrate	10.2	33.7	g
of which **Sugars**	1.1	3.6	g
Fat	10.0	33.0	g
of which **Saturated fats**	4.7	15.5	g
Dietary fibre	3.0	9.9	g
Sodium	0.3	1.1	g

3

	per 100g	per serving 330g	
Energy	659	2175	kJ
	157	518	kcal
Protein	7.8	25.7	g
Carbohydrate	10.2	33.7	g
of which **Sugars**	1.1	3.6	g
Fat	10.0	33.0	g
of which **Saturated fats**	4.7	15.5	g
Dietary fibre	3.0	9.9	g
Sodium	0.3	1.1	g

4

Carbohydrate	10.2	33.7	g
of which **Sugars**	1.1	3.6	g
Fat	10.0	33.0	g
of which **Saturated fats**	4.7	15.5	g

5

Carbohydrate	10.2	33.7	g
(**Sugars**	1.1	3.6	g)
Fat	10.0	33.0	g
(**Saturated fats**	4.7	15.5	g)

6

Carbohydrate	10.2	33.7	g
of which **Sugars**	1.1	3.6	g
Fat	10.0	33.0	g
of which **Saturated fats**	4.7	15.5	g

Numeric format (continued)

7

	per 100g	per serving 330g	
Energy	659	2175	kJ
	157	518	kcal
Protein	7.8	25.7	g
Carbohydrate of which **Sugars**	10.2	33.7	g
	1.1	3.6	g
Fat of which **Saturated fats**	10.0	33.0	g
	4.7	15.5	g
Dietary fibre	3.0	9.9	g
Sodium	0.3	1.1	g

8

	per 100g		per serving 330g	
Energy	659	kJ	2175	kJ
	157	kcal	518	kcal
Protein	7.8	g	25.7	g
Carbohydrate of which **Sugars**	10.2	g	33.7	g
	1.1	g	3.6	g
Fat of which **Saturated fats**	10.0	g	33.0	g
	4.7	g	15.5	g
Dietary fibre	3.0	g	9.9	g
Sodium	0.3	g	1.1	g

9

	per 100g	per serving 330g	
Energy	659	2175	kJ
	157	518	kcal
Protein	7.8	25.7	g
Carbohydrate of which **Sugars**	10.2	33.7	g
	1.1	3.6	g

of arrangements, since they imply the list of nutrient quantities might be summed. This implication is false: the listing shows different components, expressed in different units (kilojoules and kilocalories for energy, compared to grams for individual nutrients); additionally there is duplication of quantities in the pairs showing included nutrients (e.g. carbohydrate and sugars). Further, aligning the quantities about a decimal point would require wider columns (with quantities ranging from thousands of kilojoules to tenths of grams), and so could extend the width of the label unnecessarily.

The units for each nutrient were shown once only, to the right of both columns of quantities, so that they did not interfere with scanning across the columns. They were set in a column of their own, rather than next to the quantities in the right hand column (9), to show that they related to both columns of figures.

All the quantities were set in Univers condensed. We did not reduce the type size of the figures for the included nutrients, even though their names were slightly reduced in size, because we did not want them to be less legible, or to imply they were any less important, than the other quantities given.

Each column of quantities was headed in Univers bold condensed in the smallest size we felt possible to maintain legibility.

Verbal banding format

Verbal banding was used to eliminate the need for participants to make judgements about whether a quantity of a nutrient in a food was, for example, high or low. The quantities of nutrients in foods vary on different scales. Even if consumers know that the scales vary (for example 3g of fat in a 100g portion would be low, but 3g sodium would be high), they are unlikely to know all the scales for individual nutrients. So interpretation, in the form of verbal banding, should be helpful. We used two variations of verbal banding.

Verbal banding alongside numeric information

The banding words were placed to the right of the nutrient quantities (10). They were separated from the quantities by a slightly greater distance than the distance between the columns, to indicate that they applied to the whole row – rather than just to the information in, say, the right hand column. The banding words were set in Univers italic condensed, in order to distinguish them from other textual information. We rejected setting them in bold (too distinct) or roman (not distinct enough) (11). We prepared a version in which the banding was placed to the left of the nutrient name (12), following comments from participants in the discussion group/questionnaire study, that it was difficult to scan across two columns of numbers to banding words. However, we felt that the backward and forward scanning that this format seemed to invite made it difficult to use, especially in comparisons across foods.

Verbal banding in separate box

Banding in separate boxes could be useful where only certain nutrients are to be banded, since it less likely to appear that the banding word for a nutrient has been omitted by accident, or deliberately excluded. It could also be used where there is not enough horizontal space to include banding alongside numeric information. In our studies we used verbal banding in a separate box for comparisons with graphic formats which, because of space, had to be displayed separately from numeric information.

134

Banded Format

10

	per 100g	per serving 330g		
Energy	659	2175	kJ	
	157	518	kcal	
Protein	7.8	25.7	g	*high*
Carbohydrate	10.2	33.7	g	*low*
of which **Sugars**	1.1	3.6	g	*low*
Fat	10.0	33.0	g	*high*
of which **Saturated fats**	4.7	15.5	g	*high*
Dietary fibre	3.0	9.9	g	*high*
Sodium	0.3	1.1	g	*high*

11

Protein	7.8	25.7	g	**high**
Carbohydrate	10.2	33.7	g	**low**
of which **Sugars**	1.1	3.6	g	**low**
Fat	10.0	33.0	g	**high**
of which **Saturated fats**	4.7	15.5	g	**high**

Protein	7.8	25.7	g	*high*
Carbohydrate	10.2	33.7	g	*low*
of which **Sugars**	1.1	3.6	g	*low*
Fat	10.0	33.0	g	*high*
of which **Saturated fats**	4.7	15.5	g	*high*

12

		per 100g	per serving 330g	
	Energy	659	2175	kJ
		157	518	kcal
high	**Protein**	7.8	25.7	g
low	**Carbohydrate**	10.2	33.7	g
low	of which **Sugars**	1.1	3.6	g
high	**Fat**	10.0	33.0	g
high	of which **Saturated fats**	4.7	15.5	g
high	**Dietary fibre**	3.0	9.9	g
high	**Sodium**	0.3	1.1	g

13

	per 100g	per serving 330g	
Energy	659	2175	kJ
	157	518	kcal
Protein	7.8	25.7	g
Carbohydrate	10.2	33.7	g
of which **Sugars**	1.1	3.6	g
Fat	10.0	33.0	g
of which **Saturated fats**	4.7	15.5	g
Dietary fibre	3.0	9.9	g
Sodium	0.3	1.1	g

Protein	*high*
Carbohydrate	*low*
Sugars	*low*
Fat	*high*
Saturated fats	*high*
Dietary fibre	*high*
Sodium	*high*

The format of the banding display resembled the format of the numeric information (*13*). We used Univers bold condensed for the nutrient names, and Univers italic condensed for the banding words. Horizontal rules were used to assist horizontal scanning and to imply inclusion relationships. The included nutrients were still shown in a slightly smaller type size, but the wording 'of which' was omitted. We experimented with positioning the banding box next to the numeric information (both with and without the nutrients in both boxes aligned), but felt that placing the banding beneath the numeric information made the display clearest.

Graphic representations of banding

Banding not only interprets nutrient quantities, it also makes it easier to represent nutrient levels graphically. In our discussion group/ questionnaire study we tried out bar chart representations of nutrient quantities without banding. Trying to represent all nutrients on the same scale meant that nutrients that are present in relatively small quantities in foods, such as dietary fibre, or sodium, had very small bars (14). Using more than one scale for the nutrients would have made the displays complex. Banding makes it possible to present different nutrients on a single scale. It also limits the number of points on the scale (for example, to three, where banding is 'high, medium, or low'). So the information is easier to handle than the infinite variation on a scale that would be possible with unbanded representations of nutrient quantities.

Bar charts (direct mapping of banding levels)

The banding levels (high, medium, low etc.) were mapped directly on to different length bars on the charts. The bar charts we had used in our discussion group/questionnaire study had been presented as an addition to numeric information panels (14). Participants said they found it difficult to scan across columns of numeric information from the nutrient names to the relevant bar. So we decided that in this study we would present bar charts separately from numeric information.

Presenting bar charts separately from numeric information relaxed the constraint of presenting them as horizontal bars. We produced trial vertical bar charts (15), but rejected them for several reasons: in order to fit them within a confined space, the type size for the nutrient names had to be reduced; nutrient names had to be presented on a slant so that they could be related to individual columns, and yet be relatively easy to read; the banding words could only be linked to individual columns by making them so small they would not be legible if presented horizontally, although there was an alternative of taking banding words outside the columns (16); vertical bar charts seemed, unlike horizontal bar charts, to give an overall impression of representing a trend (perhaps because they are frequently used to represent statistical trends).

We reverted to horizontal bar charts. The nutrient names were listed, ranged right, and a bar with length representing the nutrient level placed to the right of the nutrient name (17). Note that we did not try to represent the inclusion relationships between carbohydrate and sugars, fat and saturated fats.

Although we wanted to keep the bar displays as compact as possible, we needed to be sure that their horizontal scale made the difference between banding levels clearly visible. We drafted the charts with the most detailed banding system to be used in the studies, the four-level banding system (high, medium-high, medium-low, low), in order to check that the differences in bar length for each banding level were clear.

We experimented with shading the bars with different tints, according to the nutrient level they represented, but rejected this since it meant that some bars were emphasised inappropriately (18).

We decided to place banding words within each bar rather than outside the bars (19). Although listing banding words outside the bars gave an overall impression of neatness, it was unlikely that users would ever

Graphic representations of banding

14

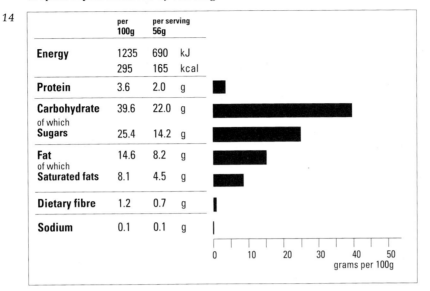

	per 100g	per serving 56g	
Energy	1235	690	kJ
	295	165	kcal
Protein	3.6	2.0	g
Carbohydrate of which	39.6	22.0	g
Sugars	25.4	14.2	g
Fat of which	14.6	8.2	g
Saturated fats	8.1	4.5	g
Dietary fibre	1.2	0.7	g
Sodium	0.1	0.1	g

15

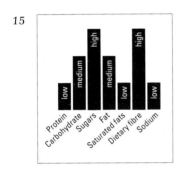

16

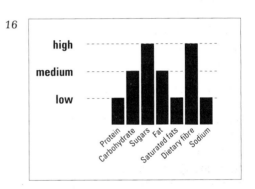

17

	per 100g	per serving 56g	
Energy	1235	690	kJ
	295	165	kcal
Protein	3.6	2.0	g
Carbohydrate of which	39.6	22.0	g
Sugars	25.4	14.2	g
Fat of which	14.6	8.2	g
Saturated fats	8.1	4.5	g
Dietary fibre	1.2	0.7	g
Sodium	0.1	0.1	g

18

Protein — low
Carbohydrate — medium
Sugars — high
Fat — medium
Saturated fats — high
Dietary fibre — low
Sodium — low

19

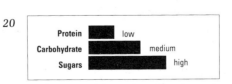

Protein — low
Carbohydrate — medium
Sugars — high

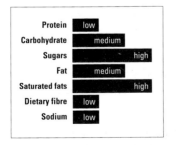

Protein — low
Carbohydrate — medium
Sugars — high
Fat — medium
Saturated fats — high
Dietary fibre — low
Sodium — low

20

Protein — low
Carbohydrate — medium
Sugars — high

Note that the banding panels illustrated in 15-20 are to be used as supplements to numeric listings (as they are shown in 17).

137

Graphic representations of banding (continued)

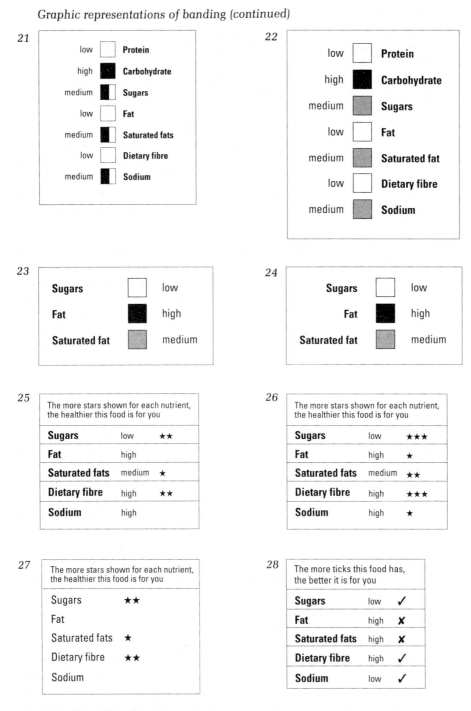

Note that all the banding panels shown on this page are to be used as supplements to numeric listings.

need to scan down a list of banding levels on its own. Placing banding words inside the bars reduced the amount of scanning needed to get from nutrient name, to bar representation, to banding word. An alternative might have been to place banding words directly to the right of each bar, rather than in a separate list (20). But this had the effect of extending the bar lengths, by different amounts, according to the lengths of the banding words.

We compared banding words presented in dark type (in this case, black) against a relatively light background formed by the bar, with banding words presented in light type (in this case, white), reversed out of a dark background. We preferred light type on a dark background, since we felt the dark bars emphasised differences in banding levels. However different production constraints might suggest different solutions: for example, if only relatively coarse definition was available, it might be wiser to put banding words outside bars; if the background colour of the package was dark, it might be better to use dark type within light bars.

Shaded boxes (direct mapping of banding levels)

Shaded boxes gave a more compact way of representing banding levels than bar charts. They could be used like a condensed bar chart with different amounts of the box shaded for different nutrient levels (21). We rejected this possibility since it seemed to offer no advantage over bar charts, other than its being compact, and since it might have been difficult to read.

We tested boxes shaded with different tints (22). We felt they should be limited to just a three-level banding system (represented by black, grey, and white boxes), because of the difficulty of ensuring that shades of grey produced in printing are distinct enough for intermediate band-ings, such as medium-high and medium-low, to be represented clearly. Even though making the distinction for single packages would be possible, ensuring comparability of greys across different packages, printed with different technologies would be very difficult.

The main difficulty in presenting shaded boxes was to find an arrange-ment of nutrient name, banding level word and shaded box that made linking the different pieces of information as easy as possible. Because the nutrient names differed considerably in length, ranging the nutrient names left, and aligning the boxes in a column to their right meant that some boxes were a long way from the relevant nutrient names (23). Ranging the nutrient names right next to a column of boxes (similar to the bar chart format) also did not give a clear enough association between box and name (24). So we put the boxes before the nutrient names, aligning the names left (22). The banding words, which served as prompts to the graphic representation, rather than the main source of banding information, were placed to the left of the filled boxes, in a normal weight, so that they had less prominence than the bold nutrient names.

Star system (evaluative mapping of banding levels)

Stars were used to indicate whether levels of nutrients indicated by banding were levels considered beneficial or detrimental to health. So nutrients such as sugars, fat, saturated fat, and sodium received more stars, the lower their level; nutrients such as fibre received more stars, the higher its level. In this way, the graphic representation gave an evaluation, rather than a direct mapping, of banding. As explained in the *Materials* section of Study 3a it would have been impossible to give carbohydrate an evaluative banding, so it was excluded from the star system, and the system was limited just to the nutrients considered to be important for health issues (protein was excluded too).

Since the evaluative system is not immediately obvious, we decided to head the system with a caption explaining that the more stars shown for nutrients the better (25). We experimented with mapping each nutrient level to a number of stars (so, for example, fat had three stars if it was low, two stars if it was medium, and one star if it was high). This produced ambiguous displays in which all nutrients received at least one star, even if their levels were considered detrimental to health (26). So we did not use stars for high levels of sugars, fat, saturated fat, and sodium, or low levels of fibre. One star was used for medium bandings of all the nutrients. Two stars were used when sugars, fat, saturated fat, or sodium were low, or when fibre was high. (The same principle could be used for a four-level banding system, with a range from no stars to three stars).

We presented the stars in a format similar to verbal banding, although, as with bar charts, we no longer showed the inclusion relationship between fat and saturated fats. In some cases stars were to be presented together with verbal banding, in other cases they were to be presented without verbal banding (27). When they were presented with verbal banding (25) we reduced the prominence of the banding words compared to the original verbal banding format (13), by setting them in upright type at a smaller size.

When stars were presented with words we used horizontal rules between nutrients to help users scan from nutrient, to banding word, to stars. But when stars were presented without banding words we found that rules tended to make the displays appear cluttered, and seemed less necessary for scanning (27). We also made the nutrient names less emphatic (reducing them to a normal weight, rather than bold).

We felt that a more direct way of representing evaluations might be to use ticks and crosses (28). However, this could only work for banding systems where it can be agreed whether intermediate levels are beneficial or detrimental to health. For example, for a three band system, it would need to be agreed whether 'medium' in fat merited a tick or a cross. We tried out this format with a two-band system (high, low) for the nutrients fat, saturated fat, sugars, sodium and fibre, where the decisions are relatively clear. We did not test the format, since its practical difficulties meant it would be unlikely to be adopted.

Presenting dietary reference values (DRVs)

Numeric DRV listing

Numeric listings of DRVs had to be positioned alongside numeric listings of nutrient quantities to allow comparisons to be made. In our standard numeric listing, the quantities of nutrients per serving were in the right-hand column, so we put the DRVs one column further to the right, moving the units (kJ, Kcal and g) to the far right (29). The listing was headed with 'recommended daily intake'.

In order to make sure people did not mistake the DRVs for the nutrient levels of the food we put a tint over them. We aimed for as light a tint as possible, to mark the different status of the reference values without obscuring the values themselves. Care would be needed in printing to ensure that legibility of the numerals was maintained.

Percentage DRV listing

Percentage DRV listings gave the percentage of the DRV for each nutrient provided by a serving of the food (30). We tested separate listings of percentage DRVs, but preferred the values listed next to numeric information. The percentages were placed at the far right of the numeric table with the heading 'percentage of recommended daily intake (per serving)'.

Graphic DRV listing

Here the percentage reference listing was represented as a number of filled circles on a scale of ten circles (31). In our discussion group/

Dietary reference values

29

	per 100g	per serving 340g	recommended daily intake*	
Energy	921	3130	9450	kJ
	221	750	2400	kcal
Protein	12.5	45.0	75	g
Carbohydrate of which	11.0	37.0	345	g
Sugars	6.9	23.5	72	g
Fat of which	14.4	49.0	80	g
Saturated fats	7.3	24.8	27	g
Dietary fibre	4.5	15.0	30	g
Sodium	0.4	1.3	2	g
*recommended daily intake based on a 2400 kcal per day diet				

30

	per 100g	per serving 340g		percentage of recommended daily intake* (per serving)
Energy	921	3130	kJ	
	221	750	kcal	
Protein	12.5	45.0	g	60%
Carbohydrate of which	11.0	37.0	g	11%
Sugars	6.9	23.5	g	33%
Fat of which	14.4	49.0	g	61%
Saturated fats	7.3	24.8	g	92%
Dietary fibre	4.5	15.0	g	50%
Sodium	0.4	1.3	g	65%
*recommended daily intake based on a 2400 kcal per day diet				

31

	per 100g	per serving 340g	
Energy	921	3130	kJ
	221	750	kcal
Protein	12.5	45.0	g
Carbohydrate of which	11.0	37.0	g
Sugars	6.9	23.5	g
Fat of which	14.4	49.0	g
Saturated fats	7.3	24.8	g
Dietary fibre	4.5	15.0	g
Sodium	0.4	1.3	g

	recommended daily intake*	
Protein	60%	●●●●●●○○○○
Carbohydrate	11%	●○○○○○○○○○
Sugars	33%	●●●○○○○○○○
Fat	61%	●●●●●●○○○○
Saturated fats	92%	●●●●●●●●●○
Dietary fibre	50%	●●●●●○○○○○
Sodium	65%	●●●●●●●○○○
* the black circles show the percentage of recommended daily intake per serving (based on a 2400 kcal per day diet)		

questionnaire study we had tried out a bar chart system giving analogue representations of percentages (*32*), but participants thought they would find it difficult to use. As with the unbanded representations of nutrient quantities, discussed above it could be difficult to judge foods on the basis of bars of different length, apart from extremes at very high or very low levels. We felt that changing the bars to a set of ten units, that could be counted up, if necessary, would make the system easier to use. In mapping the percentages on to the circles we rounded them up or down to the nearest ten per cent.

Dietary reference values (continued)

32

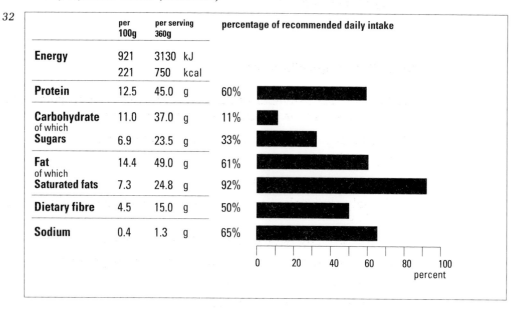

	per 100g	per serving 360g		percentage of recommended daily intake
Energy	921	3130	kJ	
	221	750	kcal	
Protein	12.5	45.0	g	60%
Carbohydrate of which	11.0	37.0	g	11%
Sugars	6.9	23.5	g	33%
Fat of which	14.4	49.0	g	61%
Saturated fats	7.3	24.8	g	92%
Dietary fibre	4.5	15.0	g	50%
Sodium	0.4	1.3	g	65%

The filled circles were shown as a separate panel from the numeric information (because of the scanning problems discussed under bar charts). The nutrient names were ranged left, with the percentage in numerals listed in a column to their right, the filled circles acting as a reinforcement. We decided that horizontal rules were necessary to help users scan from name, to percentage, to filled circles. We used horizontal rules selectively, to show the inclusion relationships between carbohydrate and sugars, fat and saturated fat.

The circles were spanned by a heading saying 'recommended daily intake', with a pointer to a footnote explaining the basis of the reference values, and the significance of the filled circles.

Appendix 2

Recruitment questionnaire

1. Are you the principal decision maker in your household about shopping, and the principal shopper for your household?

2. What recommendations can you think of for healthy eating?

   Check for:

   A   change to reduced fat milk

   B   substitute polyunsaturated margarine for butter or other animal fats

   C   reduce fatty red meats, substitute chicken/fish

   D   reduce overall consumption of full fat dairy products

   E   reduce salt (sodium) consumption

   F   increase amount of fibre eaten

   G   other (record)

3. How would you rate your interest in cooking, with 1 being the lowest rating and 7 the highest? (Show card with scale of 1-7)

4. How would your rate your interest in healthy eating? (Show card with scale of 1-7)

5. Here is a sample of nutrition information which you find on food packaging. How often would you say you look at this type of information? (Show card with nutrition label)

   Never   Occasionally   Frequently   Always

6. Do you look at this information when shopping?

7. Do you look at this information when planning meals at home?

8. When thinking about weights used on packaging labels, how do you feel about the use of grams?

   Quite happy

   Reasonably happy

   Not very happy

   Not at all happy

9. Occupation of head of household

10. Own occupation (if not head of household)

11. Age

12. Availability to attend testing session and details of name, address, phone number.

Appendix 3

Introductory summary (with dietary reference values)

Note that the final paragraph, including quantitative information, (marked here with a bar) was excluded for participants seeing the version of the summary without numeric information in Study 1, and this shortened version was used to introduce participants to other studies.

A varied diet that includes starchy foods (bread, potatoes, pasta and rice), dairy products (milk, cheese and yoghurt), meat, fish, eggs and beans, and vegetables and fruit should provide the range of nutrients we need. Most of us enjoy sweet foods and savoury snacks. If we eat these only occasionally, and not as an alternative to other, healthier foods, we can include them in our diet.

A first step for anyone who is overweight should be to lose weight by reducing the number of calories they eat. Whether we are trying to reduce our weight or not we should avoid too much fat, which is high in calories. We should cut down especially on saturated fat, which increases the level of cholesterol in the blood, increasing the risk of heart disease. We should also reduce sugar, which contains calories, but no other nutrients. And we should cut down on salt, because it can lead to high blood pressure, which can cause heart and kidney disease and strokes. We should eat foods that are rich in starch and fibre – fibre in cereals helps prevent constipation, while fibre in fruit, pulses and vegetables helps reduce the level of cholesterol in the blood, and so prevents heart disease.

Nutrition labelling can be helpful in decisions about foods to buy and eat. The amount of starch in a food is shown in the figure for carbohydrate. Both sugars and starch are carbohydrates. The amount of sugar is often listed in addition to the amount of carbohydrate.

Carbohydrate	33.0g
of which Sugars	9.0g

So the amount of starch is what remains after the sugars have been subtracted from the total amount of carbohydrate. You need to check that foods that are high in carbohydrate aren't high simply because they contain a lot of sugar.

Similarly, nutrition labelling usually gives a total figure for fat, and a separate figure for saturated fat.

Fat	9.0g
of which Saturated fat	5.5g

You should avoid a food that is high in fat if a high proportion of that fat is saturated fat.

Nutrition labelling also gives figures for sodium (the part of salt that can lead to high blood pressure) and for dietary fibre.

Within a single day, most adults who carry out normal daily activities (but not strenuous physical work) need about 2,400 calories. It's worth bearing in mind the rough quantities of different nutrients that should make up these calories:

Protein	75g
Carbohydrate	345g
Sugars	72g
Fat	80g
Saturated fat	27g
Dietary fibre	30g
Sodium	2g

Appendix 4

Nutrition bandings and dietary reference values used in studies

Nutrition bandings

The bandings are those devised by the Coronary Prevention Group (1990).* Note that dietary targets have been revised (by the 1991 COMA report)† since these banding levels were devised, and banding levels may be adjusted accordingly.

Nutrient	Dietary target	Band low	medium-low	medium-high	high
Protein	12.5% energy	<6.25	6.25-12.49	12.5-18.75	>18.75
Carbohydrate	57.5% energy	<28.75	28.75-57.49	57.5-86.25	>86.25
Sugars	12% energy	<6.0	6.0-11.9	12.0-18.0	>18.0
Fat	30% energy	<15.0	15.0-29.9	30.0-45.0	>45.0
Saturated fats	10% energy	<5.0	5.0-9.9	10.0-15.0	>15.0
Dietary fibre	30g/10MJ	<15.0	15.0-29.9	30.0-45.0	>45.0
Sodium	2g/10MJ	<1.0	1.0-1.9	2.0-3.0	>3.0

In some of the studies, only three levels of banding were used: high, medium, and low, in which case the medium band was the combined medium-high and medium-low bands shown here.

Dietary reference values

Dietary reference values were derived by expressing the dietary target for a 2,400kcal (10MJ) per day diet in grams. Note that these values pre-date the 1991 COMA report.

	g/10MJ g/day
Protein	75
Carbohydrate	345
Sugars	72
Fat	80
Saturated fats	27
Dietary Fibre	30
Sodium	2

---

* Coronary Prevention Group (1990). *Nutrition Banding.* London: The Coronary Prevention Group.

† Committee on Medical Aspects of Food Policy (1991) *Dietary reference values for food energy and nutrients for the United Kingdom.* London: HMSO.

146

References

Charny, M. and Lewis, P.A. (1987). Does health knowledge affect eating habits. *Health Education Journal*, 46, 172-6.

Committee on Medical Aspects of Food Policy (1991). *Dietary reference values for food energy and nutrients for the United Kingdom*. London: HMSO.

Coronary Prevention Group (1990). *Nutrition Banding*. London: The Coronary Prevention Group.

Commission of the European Communities (1990). *Council Directive of 24 September 1990 on nutrition labelling for foodstuffs*. 90/496/EEC.

Levy, A.S. and Schucker, R.E. (1991). *An experimental evaluation of nutrition label formats: performance and preference*. Presentation to National Food Processors Association: Scientific Forum, Chicago, Illinois.

Macdonald-Ross, M. (1977). How numbers are shown: a review of research on the presentation of quantitative data in texts. *Audio-visual Communication Review*, 25, 359-409.

Ministry of Agriculture Fisheries and Food, Department of Health, Health Education Authority (1991). *Eight Guidelines for a Healthy Diet*. London: Food Sense.

Tate, J. and Cade, J. (1990). Public knowledge of dietary fat and coronary heart disease. *Health Education Journal*, 49, 32-5.

Williams, C, and Poulter, J. (1991). Formative evaluation of a workplace menu-labelling scheme. *Journal of Human Nutrition and Dietetics*, 4, 251-262.

Printed in the United Kingdom for HMSO
Dd294090 3/92 C15 G531 10170